# A Tour of the Senses

# A Tour of the Senses

## HOW YOUR BRAIN INTERPRETS THE WORLD

John M. Henshaw

THE JOHNS HOPKINS UNIVERSITY PRESS · BALTIMORE

The Johns Hopkins University Press
2715 North Charles Street
Baltimore, Maryland 21218-4363
www.press.jhu.edu

Library of Congress Cataloging-in-Publication Data
Henshaw, John M.
A tour of the senses : how your brain interprets the world / John M. Henshaw.
p. cm.
Includes bibliographical references and index.
ISBN-13: 978-1-4214-0436-3 (hardcover : alk. paper)
ISBN-10: 1-4214-0436-2 (hardcover : alk. paper)
1. Senses and sensation.   2.  Perception.   I.  Title.
BF233.H46 2012
152.1—dc23      2011021306

A catalog record for this book is available from the British Library.

*Special discounts are available for bulk purchases of this book.*
*For more information, please contact Special Sales at 410-516-6936 or*
*specialsales@press.jhu.edu.*

The Johns Hopkins University Press uses environmentally friendly book materials,
including recycled text paper that is composed of at least 30 percent
post-consumer waste, whenever possible.

*To Dad • For making sense of it all*

# Contents

# Acknowledgments

I wish to thank the staff of the Johns Hopkins University Press for their help at every stage of this book. My students at the University of Tulsa, the most important component of my professional life, deserve special mention for having inspired this book and for providing many of its anecdotes.

# Acknowledgments

# A Tour of the Senses

# Introduction

## If You Are Lucky

If you are lucky, this is what you were born with: two eyes, two ears, a nose, a tongue, a balance mechanism in your inner ear, and a layer of skin brimming over with all kinds of sensors. If you are lucky, all are in good working order and correctly wired to your brain, which is itself continually improving its ability to process and interpret the flood of information those instruments provide.

The eyes are about halfway from the bottom of the chin to the top of the skull. Lidded and lashed for protection, these superb optical instruments distinguish millions of colors, instantly recognize faces, function in conditions varying from near darkness to intense brightness, and sort out, unaided, tiny differences between particles much smaller than a grain of sand.

On either side of the head, at about the same level as the eyes, are the ears. The exterior parts, somewhat comical-looking and formed of cartilage, aid and protect the marvelous auditory instruments inside. The ears allow the brain to distinguish and interpret sound waves whose pressures and frequencies vary over astonishing ranges.

Centered below the eyes is the nose, and just below that the tongue, sheltered inside the mouth. Often working in concert, these two instruments sort molecules based on their smell and taste. The nose can distinguish about ten thousand different odors, protecting us from danger and enhancing our quality of life in countless ways. The tongue is at once the last sense guarding the body from poisons disguised as nourishment and an organ of immense sensual pleasure.

In the inner ear, next to the hearing organ, a set of fluid-filled instruments monitors the movements of the head with amazing fidelity. With no visible external parts, the vestibular system, as it is called, is surely the most underappreciated of the sense organs, although its duties could scarcely be more vital.

Last but hardly least, touch is our most extensive sense, covering as it does most of our body—wherever there is skin. At its most sensitive in the fingertips, it allows the blind to read and everyone to distinguish thousands of different textures. But that's only the beginning. Receptors in the skin and elsewhere also sense temperature, pain, and, of enormous importance, the position of the various parts of the body.

All of this superb instrumentation is useless by itself. The data gathered must be filtered, reconstructed, and interpreted, and that is the job of the brain—itself mysterious in many ways, but becoming better understood all the time.

These are our senses. There are more than five, but who's counting? Each sense organ is an instrument of surpassing grace, efficiency, and versatility that converts external stimuli into electrical signals that are acted upon by our conscious and unconscious selves, in ways we have yet to completely comprehend, through the offices of our central nervous system and our brain, the most magnificent organ of all.

From the moment we are born, and even before, we develop and grow intellectually, socially, and emotionally by bringing information to our brain through our senses. And for most of our history as a species, the natural senses have been unrivaled, by any contrivance of humankind, in their ability to provide us information and knowledge. That has begun to change, as more and better sensing devices are developed. Some, such as cochlear implants, help restore a missing or deficient sensory ability inside the human body. Others—and here the list is very long indeed—work independently, such as the automated face-recognition systems to identify terrorists and other criminals.

Advances of these kinds will continue to change our lives. There are the obvious benefits to persons with disabilities, to medical professionals, and to law enforcement personnel. But this revolution in sensory instrumentation also has consequences for human development and for education. Why, for example, should a child learn to draw when high-quality digital cameras are so cheap and simple to use? Why should a medical student learn to palpate (feel with his fingers) a patient's body, when so many powerful instruments are available to give him the same information and more? Why should an aspiring acoustical engineer listen to performances in a concert hall, when all he really needs to do is set up his instruments properly and let them "listen" for him? To these and similar questions, my answer is that the task of educating and protecting our senses is as important as it ever was and that we de-emphasize it at our peril. That the adult's senses are so much more refined than those of a baby is a result of education, both formal and informal. Destructive forces lurk, however, as our senses are daily assaulted by unwanted noise, omnipresent video screens, and noxious odors.

Come along with me as we take stock of the sensory gifts we were given, of what we have learned how to do, through technology, to aid or improve the senses (and how such technology is likely to change), and of where this leaves us, in terms of interacting with the world around us. Having completed my own tour of the senses, I confess to an overwhelming sense of awe at the magnificent gifts we have been given. In spite of the astonishing rate at which science continues to unravel their mysteries, what we don't know about our senses sometimes dwarfs what we do know. And the technological advances, while impressive and in some instances capable of feats we cannot accomplish unaided, have a long way to go to match our natural gifts.

## Stimulus, Sensation, and Perception

It was tempting to organize this book around the five best-known senses: vision, hearing, taste, smell, and touch. But that approach misses some important truths about the sensory process. I have concluded that a tour of the senses does not logically proceed from vision to hearing to smell, taste, and touch, but rather from stimulus to sensation to perception. Those are the names I have given the three main parts of the book.

*Stimulus.* We live in a world crammed with sensory stimuli. There's so much that it's a miracle we can bring any order to our lives out of it all. No wonder we sometimes complain of sensory overload. Electromagnetic waves rain down upon us from all directions, the air we breathe is constantly vibrating around us, and that same air is full of complex organic and inorganic molecules. There are also plenty of stimuli we humans can't sense, but which some of our fellow creatures can, such as infrared waves and magnetic fields.

*Sensation.* Long ago, creatures began to evolve various sensory abilities, various flesh-and-blood instruments. Different stimuli require different instruments. Over genetic time these instruments or organs have become highly specialized in humans and other animals, though not always in the same ways. This is sensation, and it is the beginning of everything we know.

*Perception.* The results of those sensations, the data gathered, are the beginning, but they are of little use without a lot of processing. Vast areas of the human brain are devoted to countless tasks associated with acquiring, filtering, transforming, reconstructing, integrating, and organizing the information gathered in the sensation processes. How else could you, for example, pick out and locate the faint voice of your lost child crying out amid the cacophony of a crowded train station? This is perception.

Problems sometimes arise. There are individuals for whom the stimulus-sensation-perception process breaks down in one way or another. Many of our sensory organs begin to perform poorly over time, betraying their owners well before the end of their days. Perceptual problems also afflict many people. We can fix some of these sensory and perceptual difficulties, or at least improve the situation. Some of the recent advances are truly spectacular. Other such afflictions have not yet yielded to our efforts.

How best to make use of our sensory gifts? A formal education focuses less on the senses than it once did. Formalized training of the senses has too often become the highly specialized preserve of the artist, the musician, the athlete, or the chef. The training of professionals, such as doctors, once critically depended on developing the senses as data-gathering tools. Those same professionals now devote their time to mastering the instruments that have, over time, largely replaced their senses.

It's tempting to think of the progression from stimulus to sensation to perception as beginning in the realm of physics and chemistry and then moving on to biology and physiology before finally ending up in psychology and even philosophy. There is some truth there, but things are more complicated than that, and we shall make numerous side trips on our tour. One of my favorite professors once told me that the most interesting things in science are often found in the nooks and crannies between the traditional academic disciplines. The boundaries between stimulus and sensation (physics and physiology) and between sensation and perception (physiology and psychology) are good examples of this truth. My old professor was right.

## The *Five* Senses?

Here's a conversation between a writer and a casual acquaintance.

> "I hear you're writing a book."
> "Why yes, I am."
> "What's it about?"
> "The senses."
> "The *five* senses?"
> "Well, actually . . ."

If this book does nothing else, perhaps it will convince you that there are more than five senses. Because there are. Period. Try telling someone that, though, and the person is likely to conclude you're a bit of a wacko. The idea that we have exactly five senses dates back at least to Aristotle (384–322 B.C.). In *De anima* (*Of the Soul*), Aristotle writes a great deal about the senses, at one point even providing something of a proof that there can be no sixth sense, no sense beyond sight, hearing, smell, taste, and touch. For each sense there is a sense organ, he argues, and since we only have five such organs, there can be only five senses. The logic is sound, but the physiology is found wanting, because we do indeed have organs for more than five senses. Aristotle can be forgiven, since some of those sense organs were discovered long after his time. But the number five persists, perhaps because of Aristotle, and perhaps because those five senses remain the most tangible and obvious ones.

According to Merriam-Webster, a sense organ is "a bodily structure that receives a stimulus (as heat or sound waves) and is affected in such a manner as to initiate a wave of excitation in associated sensory nerve fibers which convey specific impulses to the central nervous system where they are interpreted as corresponding sensations." Those "corresponding sensations" are the end products of our senses. Sensing something always involves converting a stimulus, such as a sound wave, into an electrical signal. That is true of our natural senses and of human-made sensors. In the human body, the electrical signals generated by the sensors are interpreted by the brain. For human-made sensors, a computer generally does the honors.

Based on those definitions and on what we know about the human body, there are clearly more than five senses. The skin, all by itself, contains different types of receptors associated with four different senses: touch, temperature, pain, and body awareness or proprioception. The latter relates to sensors that allow us to keep track of where various parts of our body are at any given time. Put your hand behind your back, and make a fist. How can you be sure you really made a fist? The proprioceptive sensors in your hand—that's how. Another underappreciated sense organ is the vestibular system, a miraculous little set of devices housed inside the skull that sense the motion of the body, and in particular, the head; the vestibular system enables us to balance ourselves. Nine would be a much better estimate of the number of senses we possess than five. Those nine are vision, hearing, taste, smell, touch, temperature, pain, balance, and body awareness.

## Technology and the Senses

Merriam-Webster will soon be faced with redefining *sense organ*. The advent of the cochlear implant has changed forever what it means to sense something, for it no longer requires in every case a "bodily organ" to do so. Humankind has been devising ways to aid the senses for centuries (eyeglasses date back to about 1300), but the cochlear implant represents the first device that actually replaces sensory receptors in the human body. Sensory receptors convert external stimuli into electrical signals. In 1973, Arthur C. Clarke wrote, "Any sufficiently advanced technology is indistinguishable from magic," and today that's how the cochlear implant seems to me—like magic. A cochlear implant is not a hearing aid. Not even a really, really good hearing aid. Devices like hearing aids and eyeglasses can only enhance or augment a sense organ. A cochlear implant actually replaces the cochlea, the magnificent organ in the inner ear where "hearing" really takes place and where the mechanical energy in sound waves is finally transformed into electrical signals.

What will be the next device of this kind, the next artificial sense organ? Perhaps it will be an artificial retina. Only time will tell. In the meantime, technology marches forward in other sense-related areas. I got a phenomenal toy, or rather, "research tool," recently: an infrared camera. It allows me to see infrared radiation, heat waves, something very few creatures on earth can do unaided. The world seen through the camera's LCD screen is vastly different from the one afforded me by my unaided eyes. Looking at a blank wall, I can see where a hot water pipe runs behind it. In a parking lot, I can tell which cars recently arrived and which ones have been sitting there all day. From the front yard, I can tell at a glance which of my house's windows are single-pane and which are of the more insulating double-pane variety.

This whole idea of replacing the senses with other things is not all that new. The ancient Greeks did not especially trust their natural senses, as Frederick Hunt describes in his book *Origins in Acoustics*. Pythagoras, of right-triangle fame, and some of his colleagues seemed to gradually lose faith in their senses as arbiters of right and wrong, true and false, and began to seek to interpret everything in mathematical terms. Heraclitus (ca. 536–470 B.C.) maintained, "The eyes are more exact witnesses than the ears" and that "the eyes and ears are bad witnesses for men, if their souls lack understanding." Anaxagoras (ca. 499–428 B.C.) put it more forcefully: "Through the weakness of the sense-perceptions, we cannot judge truth." Philolaus summed things up another way, and geeks everywhere have been rejoicing ever since: "Actually," he said, "everything that can be known has a Number; for it is impossible to grasp anything with the mind or to recognize it without this Number."

Nowadays we find that we can, through instrumentation, overcome many weaknesses of the senses. As the cochlear implant and the infrared camera indicate, instruments that hear, see, and sense in other ways are getting more sophisticated all the time. In general, such instruments provide us with numbers in abundance, and the computers they are nearly always attached to massage those numbers into a form that we can more easily interpret, such as the electronic signals from a cochlear implant, the color image on the screen of my IR camera, and the 3-D color image from an MRI.

I've never owned a boa constrictor or any other kind of snake. But ever since I got my IR camera, I've felt a certain kinship to the boa, a creature that can hunt its prey at night, aided by an innate ability to sense infrared radiation. The animal kingdom is home to a rich diversity of sensory abilities.

# The Bat, the Narwhal, and the Bee

Animals' senses can differ from ours in magnitude or in kind. Everyone knows, for example, that dogs have a much more powerful sense of smell than humans. Dogs can be trained to sniff out drugs and bombs, feats that seem to be out of the question for humans. But this is merely a magnitude difference. Dogs and humans both possess the sense of smell; it's just that a dog's sense is much more acute. There are plenty of other magnitude sensory differences, and in those cases too, humans often come up short when compared with other creatures.

Then there are the differences of kind. Some of the senses found exclusively among nonhumans are the ability to sense ultraviolet or infrared waves, the ability to detect fluctuations in magnetic fields, and the ability to detect electrical fields. Echolocation, the ability that bats have to locate objects through reflected sound waves or echoes, is a sensory ability most humans do not possess. This is a difference of magnitude and not of kind, though, because echolocation is really just a highly specialized form of hearing. Consider a remarkable Californian named Ben Underwood.

Ben lost both eyes to retinal cancer at the age of three. Soon thereafter, his hearing began to allow him to sense things normally reserved for the sighted. He could tell when the car he was riding in was passing a tall building, because the echoes from the traffic noise would change. He could hear that difference, and over time he learned what it meant. By the time Ben was seven years old, he had taken his sense of hearing to a different level. He developed the habit of making strange clicking noises and using the echoes to navigate his sightless world. His special form of echolocation enabled him to walk without a cane or a guide dog and even to ride a bicycle or a skateboard. Researchers at the University of California, Santa Barbara, confirmed Ben's ability to detect and differentiate small objects based on shape through echolocation. Cancer attacked Ben again when he was a teenager, and he died in January 2009, at age sixteen.

As remarkable as he was, Ben Underwood was not as proficient as a bat, a whale, or a dolphin, creatures whose echolocation skills are legendary.

Ben's startling ability to echolocate was limited by his utterly normal human hearing instruments: his ears.

The ability to detect electrical fields, called electroception, is found in aquatic creatures such as the electric eel, the great white shark, the hammerhead shark, several species of rays, and the platypus. It is believed that these animals sense high-frequency alternating currents in order to detect the muscle activity of other animals and perhaps also as an aid in navigation.

The ability to detect magnetic fields, or magnetoception, is a related sense found in pigeons, loggerhead sea turtles, lobsters, honeybees, rainbow trout, and mole rats, among others. Detecting the magnetic field of the planet equips these creatures with a sort of internal compass and serves as an aid to navigation.

Certain snakes, such as boa constrictors, appear to be the only creatures capable of sensing infrared radiation, or heat waves. This is a difference of kind, not of magnitude, because these snakes don't utilize their eyes for this purpose. They possess a separate organ, called a pit hole or pit organ, that enables them to hunt warm-bodied prey in the dark. How all this works is explained in part 2, "Sensation."

A near-sighted honeybee, if there is such a thing, wouldn't much like my eyeglasses. Never mind that they'd be a bit large for the little fellow; there is another problem: my glasses are designed to filter out ultraviolet (UV) radiation. Humans can't see UV radiation, and since it's bad for our eyes, filtering it out through eyeglasses is a good idea. Honeybees *can* see UV radiation, however.

Ultraviolet waves have higher frequencies than visible light. On the lower-frequency side of the visible spectrum lies infrared. UV waves, the ones that give you sunburns, are invisible to humans, but honeybees can see them just fine. Unlike the snakes that sense IR with their pit holes, honeybees use their eyes to sense ultraviolet. The sensors in their eyes are optimized for higher-frequency waves than those visible to human eyes. Humans see the "visible" spectrum, from low-frequency red to orange, yellow, green, blue, and finally violet on the high-frequency end. Honey-bees can't see red and thus aren't attracted to red flowers. Instead, these bee eyes see orange, yellow, green, blue, violet, and ultraviolet. Their ability to sense UV has a lot to do with which flowers they are attracted to, and when.

Echolocating bats, UV-sensing bees, and snakes that can detect heat waves in the dark are all bizarre enough from the human perspective. But when it comes to unusual sense organs in the animal kingdom, it's hard to top the strange case of the narwhal. A member of the whale family found mainly in arctic waters, the narwhal possesses a tusk of mythic proportions. A straight, slender, solitary, conical appendage with a graceful spiral on the outside, the narwhal's tusk can reach a length of nine feet. Given that the narwhal's body grows to no more than fifteen feet, the tusk is truly immense. It has been described as resembling a cross between a corkscrew and a jousting lance, and what its purpose could be is one of the oldest mysteries in the world of natural history. The answer is even stranger than many of the outrageous myths that have arisen over the centuries: It has been suggested that the tusk is used to poke holes in the ice (the narwhal must surface to breath) or that it is a weapon (the aforementioned jousting lance) used for fishing, hunting, defending its young, establishing dominance in a group, even for bashing holes in the bottoms of ships, or some combination of the above. Other theories have held that the tusk is a sound-transmitting device or a radiator used to cool the animal's body.

Humankind has ruminated on the purpose of the narwhal's tusk since at least the year A.D. 1000. For centuries, tusks were collected and fobbed off as "unicorn horns" in one of the longest-running and most profitable scams in history.

There are no unicorns, and none of the above explanations have any truth. It took a practicing dentist with an adventuresome streak to begin to discover the real story behind the narwhal's tusk. It is a tooth that has evolved into what appears to be primarily a sense organ. Dr. Martin

Nweeia and his research team discovered that the surface of the narwhal's tusk is covered with millions of nerve endings. The tusk is the left front tooth of the narwhal. The right front tooth is generally only about a foot long and remains inside the animal's body. The long tooth or tusk is most common in the male, although some females do exhibit a tusk. Males with two long tusks are rare but have been observed.

Dr. Nweeia established that the nerve endings at the surface of the tusk are capable of detecting changes in the saltiness of ocean water. This information could warn the animal that the water at the ocean surface is beginning to freeze. When salt water begins to freeze, almost all the salt remains in the liquid water, and not in the ice. Thus, the water in the vicinity of newly formed ice becomes saltier.

It is also suspected that the nerve endings on the narwhal's tusk can sense temperature changes, pressure changes, and perhaps other things. It could be that the narwhal's tusk is something like a versatile weather station, monitoring not only changes in salt concentration in the water but also air temperatures and pressures, since the narwhal has the habit of floating on the ocean's surface, its head angled so that its tusk points straight up, like an antenna.

## Seeing Is Believing

Not all eyes are created equal. Eagles have better long-distance vision than humans, cats see better at night, and flies have segmented eyes to see in multiple directions at once. Even among people, there are differences, including obvious ones such as color blindness and various correctible deficiencies such as myopia or astigmatism. Beyond all that, however, lie the most interesting differences of all.

The term *genius* is overused, but in the case of artists like Leonardo or Rembrandt, it is richly deserved. Somewhat more recently, baseball fans marveled at the eyes of famed slugger Ted Williams. And in the 2009 Super Bowl, fans were similarly amazed by the feats of Larry Fitzgerald, the pass-catching whiz of the Arizona Cardinals, who seems to have eyes in the back of his head. Does the artistic or athletic genius of these folks come from their eyes, their brains, their training, or some combination of all three?

Teachers, if they are honest, will admit they learn at least as much from their students as they manage to teach them. This book has its roots in an innocent little incident that took place a dozen years ago, when one of my students redefined for me something I thought I already knew, and that is just how important the senses are in education. As a graduate student, I had been interested in the creative aspects of the engineering design process—how it is that engineers go from a clean sheet of paper, or a blank computer screen, to a fully realized creation such as a car, a photocopier, or an MP3 player. In my research, I came across a book by Betty Edwards called *Drawing on the Right Side of the Brain*.

Edwards's book, among other things, is a how-to guide to drawing. She contends that the reason most of us are lousy at drawing is not so much a lack of artistic ability, but rather that we have so little formal training, especially vision training. Most of us in the United States receive no formal instruction in drawing skills beyond about the third grade. Ask an adult with a third-grade drawing education to sketch someone's portrait, and you're likely to get a result that appears to have been created by a third-grader. Our drawing skills are frozen in time, back in the third grade.

But people assume their drawing shortcomings arise because they have no "artistic ability," and not, as Edwards contends, as a result of a lack of training. Ask someone to make a pencil sketch of something, and the person is likely to demur, saying, "I'm sorry, but I just don't have any artistic talent." On the other hand, if you asked nearly any high school graduate in America to write a short business letter, he would almost certainly be able to produce a reasonable result. The letter would not remind you of a third-grader's effort, nor would its writer be likely to offer an excuse such as "I just don't have any writing talent."

Edwards contends that we look at these two skills, making a simple drawing and writing a simple letter, so differently because of the training we have received. We seem to believe that because we cannot draw like

Picasso, we have "no artistic talent." But writing is different. Just because someone will never write like Nabokov, he would never say, "I have no writing talent." Writing is important in daily life, and our formal education, through high school and beyond, emphasizes it. If we can get past the idea that only artists can draw, Edwards says, we can all learn to be competent at this skill. She proceeds to pick up our drawing education from where it left off back in the third grade.

And this is where the bit about the senses and education, and my own re-education, comes back into the story. One of the most important lessons in Edwards's book is that in order to be able to draw something, you have to be able to see it—really see it. She uses various tricks to get us to see familiar things from new and different perspectives. For example, if you want to learn to sketch a portrait, try this: in a magazine, find a photograph of a face you'd like to draw. Turn the photo upside down, and start drawing. Since we aren't used to looking at faces upside down, our brains are deprived of any notions about how such things should look, and our eyes are forced to consider the image in a new way.

From an upside-down photo, we are much more likely to draw what we *really* see, rather than what our brains *think* we should see. If your refrigerator happens to be adorned with the artwork of any children, you probably have a good example of this. Look at a third-grader's drawing of someone's face. Where are the eyes? Chances are, they have been placed far too high up in the head. In reality, our eyes are located about halfway between the top of the head and the bottom of the chin. If you don't believe me, flip a photo upside down, and see for yourself. But a third-grader doesn't draw a face that way. He places the eyes way up near the top of the head, in the middle of where the forehead should be. This is because most of the other interesting stuff in a face is located from the eyes down, namely the nose and the mouth. This is what our brains are interested in. So when we draw a face, we don't really *look* at the face so much as we just put down on paper what our brain tells us a face should look like. In drawing, the information should flow from the eyes to the brain to the hand. But for those of us who lack the training, the brain takes over and more or less cuts the eyes out of the process.

I thought I understood all this by the time I became a professor of engineering. As a result, I occasionally required my students to make pencil drawings of some of the things we'd done in the lab, in order to force them to *really* look at whatever it was I wanted them to see. I still do this, and it becomes less popular each year, especially given that nearly every student now carries a digital camera, built in to his mobile phone. "Tell me again why I can't just take a picture of this?" is a question I often get in the laboratory. It's like this, I say, and I recount the Betty Edwards story. The students grumble a little, but they do what I ask. Engineers are great that way. And I still think it's worth the effort; otherwise I'd have stopped doing it long ago.

It was long ago when a nontraditional (that is, older-than-usual) college student taught me, through his own example, just how poorly I'd learned Betty Edwards's lessons myself. This guy had gone to work as a commercial artist, painting signs and designing company logos, straight out of high school and had made his living that way for a dozen years. At that point he packed up his paintbrushes and began his college education, studying engineering. He was a marvelous student, as nontraditional students often are. It was his junior year, and I had him in a lab where all the students were assigned to make drawings of these pieces of steel we'd broken during a lab experiment. This particular experiment is one with which I was more than familiar. (It's called the Charpy impact test, for all you engineers out there.) I'd first done it as a student, then many times during my engineering career in industry, and then once every year after that as an engineering professor. During that time, and having become a disciple of Betty Edwards, I'd looked at those little pieces of broken steel so many times— upside down, right side up, and inside out—that I would have bet my house nobody could teach me anything new about them.

Until my artist-cum-student came along. His sketches, when they arrived with his lab report, were breathtaking. First of all, he filled an entire page with each sketch of these little pieces of steel not more than a few inches long. Another Betty Edwards lesson is to make your drawings large. This forces your eyes to see more detail. Second, his sketches were beautiful. The typical engineering student has "no artistic talent" (or rather he has a third-grade art education), and I've grown accustomed to this. When I evaluate their pencil sketches, I'm looking for evidence that they really made an effort to see and record the fine details in the broken steel, and not much more.

Finally, my artist-student's sketches contained a few surprises, or should I say revelations? In his drawings, he'd penciled in details I had never been aware of. Certain kinds of steel had deformed and fractured in certain subtle ways that I had never noticed before. No matter that I'd looked at these things dozens of times. Or perhaps I'd never really looked at them beyond the first few times. My brain had evidently started telling my eyes to take the day off. "We've seen these things before. Nothing new here." Just like the third-grader who puts the eyes in the middle of the forehead of his portrait, I had stopped looking at these little pieces of steel. And now, this student, who had *never* seen one of these samples before, had taught me things about them that I didn't know. I almost fell off my chair when I realized this. I ran down to my lab with his sketches in hand to compare them to the real thing. Sure enough, he was right. That guy could see, and I mean really see. His brain was trained to pay attention to his eyes—full stop.

That was years ago. Since then, a handful of other students have managed to notice and include some of these same details in their sketches of the results of that experiment. It doesn't happen very often. Each year after I've graded everyone's reports, however, I do make sure to point out those little details my artist friend first showed me, while I tell them the story I've just told you.

And ever since I looked at my artist-student's sketches for the first time, I've been thinking more and more about the senses and education, or just about the senses themselves. I began to wonder about the whole process of how we become aware of the world around us, the process that proceeds from stimulus to sensation to perception. It's a fascinating, complex, exciting story that is at once well understood and filled with mysteries. As we unravel those mysteries, we learn more and more about what it means to be human. That's the story I want to tell in this book.

# Part 1 • Stimulus

We are aware when the room gets darker, or colder, or louder, or if there is a gas leak. But how many different wireless telephone call signals are passing through your body this very instant?

Things like that, whether we're aware of them or not, we'll call stimuli. Stimuli include visible light, other kinds of waves we can't see (x-rays, ultraviolet, infrared), vibrating air (within certain frequency limits and above certain threshold amplitudes), various types of molecules floating around in the air, changes in temperature, and the rate at which your body is moving in various directions. Our bodies have evolved the ability to detect these stimuli and others, through a group of processes broadly called sensation. The brain processes the results of those sensations. This is called perception. In part 1 we focus on stimuli our bodies can detect and explore some of the ones we can't. The world we know is so intimately defined by the stimuli we can sense that this has become our reality.

Our reality is not the same as that of the other animals with which we share the planet. Evolution moves in different directions, and thus we find animals that are aware of stimuli to which humans are oblivious, such as infrared waves or magnetic fields. Oblivious, that is, until we figure out how to make instruments that measure those particular stimuli. For example, my infrared camera can take photos that reveal the temperatures of things. The fine detail in those photos would surely make any boa constrictor, a snake possessing sense organs that detect infrared, green with envy.

The stimuli that humans, other animals, and human-made machines can detect can be divided into three broad categories: electromagnetic stimuli, chemical stimuli, and mechanical stimuli.

# Chapter 1 • Electromagnetic Stimuli

## The Electromagnetic Spectrum

The different types of electromagnetic waves that emanate from a doctor's x-ray machine, from your mobile telephone, from a TV remote control, from a microwave oven, or from a car's headlights are not as different as you might think. They *are* different, but they have something in common. They are all part of the electromagnetic spectrum, which extends from the most energetic waves, such as gamma rays and x-rays, to the least energetic, such as radio and TV waves, as shown in figure 1. The special names we give to various portions of the spectrum (ultraviolet, infrared, radio, and so on) represent rather arbitrary divisions, much as the boundaries between states or countries are somewhat capricious. But the behavior of the waves does vary, spectacularly, as we move from one end of the spectrum to the other, and so the names for the various regions have a certain utility. The more energetic the waves, the shorter their wavelengths are, and the greater their frequency. As frequency goes up, wavelength goes down. Mathematically, frequency is equal to the speed of light divided by the wavelength.

The simple relationship between frequency and wavelength holds for any type of waves, not just electromagnetic waves. Sound waves, which are mechanical and not electromagnetic, are a good example. For sound waves, frequency equals the speed of *sound* divided by wavelength. When

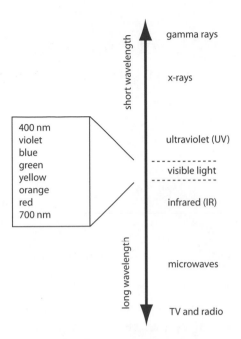

*Fig. 1.* The electromagnetic spectrum.

we discuss light and other forms of electromagnetic waves, we'll do so mostly in terms of wavelength. Later on, for sound, the discussion will generally center on frequency. This is typically how these things are discussed, but it could just as easily have been the other way around.

X-rays have wavelengths around 1 nanometer, while FM radio waves have wavelengths around 10 meters, or 10 billion times longer than x-rays. The electromagnetic spectrum, then, is enormously wide. A tiny slice of the spectrum is termed, for obvious reasons, "visible" light, for that is what we can see. The visible range of the spectrum is quite narrow, extending from a wavelength of about 400 to a wavelength of about 700 nanometers (nm). The 400–700 nm range is the one listed by the Optical Society of America and others. The spectrum of visible light varies depending on the source. Wavelengths as low as 380 nm and as high as 750 nm are sometimes reported.

There are various explanations for why human sight evolved in what we now call the visible region of the electromagnetic spectrum. One of the explanations involves the ability of different kinds of waves to travel through different types of materials. Vision almost certainly evolved first in sea-dwelling creatures. Later, when life moved onto land, those creatures brought vision with them. Whatever forms (whatever wavelengths) of electromagnetic waves the early sea creatures were sensing, those waves were transmitted to them through water.

Aquatic life exists that can sense other types of radiation besides visible light. Witness the electric eel and some species of sharks, which have the ability to detect the electrical currents emitted by the muscle movements of potential prey, or the loggerhead turtle and the rainbow trout, which possess a related ability to detect magnetic fields. The latter is no doubt useful during migration. But these are not vision. These senses involve other sense organs. The eyes evolved to detect "visible" light, over its characteristic narrow range of frequencies, and the nature of water gives a clue as to why.

Water absorbs quite effectively wide swaths of electromagnetic energy above and below the visible spectrum. Both ultraviolet (wavelengths just shorter than visible light) and infrared (just longer) are absorbed almost completely by water. Nestled in between ultraviolet and infrared is a patch of radiation extending from wavelengths of about 400 to 700 nanometers that is absorbed hardly at all by water. This can scarcely be a coincidence.

The spectrum of sunlight is shown in figure 2, which shows that it's not just visible light that the sun delivers. Significant quantities of ultraviolet and infrared waves also reach the earth's surface. About 8 percent of the total energy contained in sunlight is ultraviolet, 46 percent is visible, and 46 percent is infrared. UV and IR thus enter the ocean in huge quantities in the form of sunlight, along with visible light. But whereas the IR and UV radiation are quickly absorbed—turned into heat—by the water, the visible light passes through the water relatively freely. But not infinitely so. The deeper you go in the ocean, the less visible light there is. At depths

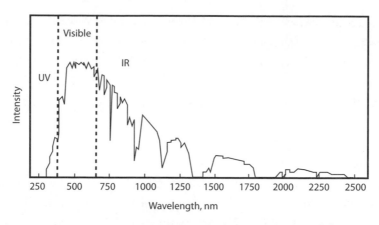

*Fig. 2.* The spectrum of sunlight at the earth's surface.

below about 2,000 meters, the darkness is so profound that creatures living down there, such as the magnapinna squid, are blind.

If you'd been a sea creature back when vision was first evolving, about the only type of electromagnetic waves that reached your body would have been what we now call visible light. Life was simpler back then, if particularly brutish and short, dominated as it was by the imperatives of finding things to eat, avoiding being eaten, and procreating. Today, vision retains great utility for each of these purposes, and it isn't hard to imagine that it could have evolved that way.

But in any event, it wasn't going to evolve in the ultraviolet or infrared ranges, because of the efficiency with which water absorbs those types of waves. You can't sense electromagnetic waves, or any other sort of stimulus, if they never get to you. We are thus stuck with "visible" light when it comes to vision.

A former student of mine swears that her father has vision in the infrared range. She claims he can see the IR beam that emanates from a TV remote control unit when you push one of the buttons. She says he has demonstrated this ability numerous times, although I have not witnessed it myself. It does seem rather incredible, since TV remote control devices typically produce IR signals with wavelengths of 980 nm or so, far longer than the 700 nm upper end of the visible range.

The human eye contains four different types of light sensors or receptors. Three of those are optimized for color vision, and the fourth is optimized for low light conditions. The three color receptors, called cones, are optimized, respectively, for short, medium, and long wavelengths within the visible spectrum. Were the human visible spectrum any wider, further receptors would be required. The width of the visible spectrum is thus defined by the capabilities of the rods and cones, a story we shall return to in part 2.

In *Empire of Light*, Sidney Perkowitz uses a musical analogy to explain just how little of the electromagnetic spectrum is occupied by visible light. The great Hermann von Helmholtz presented a similar explanation in 1867. In music, two notes are one octave apart if the frequency of one of the notes is twice that of the other. The eighty-eight keys on a piano span a frequency range from 27 to nearly 4,200 hertz,[1] or just over seven octaves. Human hearing normally extends over a range of 20 to 20,000 hertz, which is almost ten octaves. Start with 20. Double it to 40, then 80, then 160. The tenth doubling brings you to 20,480.

It isn't often done, but the electromagnetic spectrum can be expressed in terms of octaves as well. There are about thirty-seven octaves between a typical FM radio wave, on the low-frequency end of the spectrum, and a typical x-ray, on the high-frequency end. In between, human vision occupies a humble one-octave-wide slot, about ten octaves below a typical x-ray. Perkowitz notes that if human *hearing* had a range of only one octave, instead of ten, its utility would be severely limited indeed.

## Detecting Electromagnetic Waves

We can't see or otherwise sense longer or shorter wavelengths than visible light, but that hasn't stopped humankind from developing instruments

---

[1]A hertz is one cycle per second.

that detect and measure such stimuli and convert them into useful forms of information. The electromagnetic spectrum is a vast playground for the senses, both natural and synthetic. Concerning the latter, the number of different types of instruments developed by humankind for measuring and interpreting different types of electromagnetic waves is staggering.

One example is the radiograph, which measures x-rays. An x-ray machine is familiar to most of us through its medical uses, but this technology is also useful for inspecting things like pipelines and nuclear reactors. What is measured is density differences. X-rays have tiny wavelengths and are highly energetic, so energetic that they can pass right through many solid objects. But the denser the object, the more it will absorb the x-rays. Shine a beam of x-rays on an arm that contains a broken bone, and place a sheet of photographic film behind the arm. (These days, digital radiographic imaging technologies are widely employed, replacing the film— just as digital cameras have mostly replaced the ones that use film.) The density differences in the bone will be recorded in the form of a visual image on the resulting radiograph. A hairline crack, being less dense than the solid bone around it, will allow more x-rays to pass through and will show up as a thin dark line on the radiograph.

But in some ways a radiograph represents a rather low-tech application of x-rays as a synthetic sense. There are other technologies that utilize x-rays in sensory roles, such as x-ray diffraction. When any kind of waves come into contact with matter, be it solid, liquid, or gas, a variety of things can happen. Shine a flashlight on a window pane, and most of the light passes right through the window. We say it is transmitted. A small percentage of light is absorbed by the glass,[2] however, and another small portion

---

[2]The amount of light or other radiation transmitted through a material like glass is proportional to how thick the material is. A one-quarter-inch-thick pane of glass might absorb only a few percent of the energy contained in visible light. For a windowpane, this amount is negligibly small. For applications like fiber optics, where the goal is to transmit light through glass fibers for many miles, it's a different story. Back in the 1960s, researchers struggled to transmit light through glass fibers that were only a few feet long. It was discovered that tiny impurities in the glass absorbed

is reflected. The same three things happen to a piece of broccoli in a microwave oven, except that the percentages are different. A large fraction of the microwave energy is absorbed by the water molecules in the broccoli, thus heating the food, while the rest is either transmitted or reflected. Now think of a parabolic satellite dish. The long-wavelength TV signals that arrive at the dish from the satellite are reflected by the dish onto the detector and from there transmitted to your television, but some of the incoming waves are either transmitted or absorbed.

Electromagnetic waves can be transmitted, absorbed, or reflected by a material. But there is another possibility: diffraction. *Diffraction* is a general term referring to the bending or redirection of waves that occurs when the waves interact with various patterns of obstacles. Any kind of wave can undergo diffraction. As ocean waves approach the shore, they can be diffracted by rocks and jetties. Sound waves diffract around corners, allowing us to hear someone talking in another room. Visible-light waves can be diffracted as well. The rings that are sometimes seen around the sun or the moon occur when the sunlight or moonlight is diffracted by thin clouds in the atmosphere. (The familiar rainbow is mainly a product of *refraction*, which is the bending of light waves as they are transmitted, in this case from air to water and back to air again.)

X-rays can be diffracted as well. Diffraction typically occurs when the wavelength of the radiation involved is about the same size as the spacing between the obstacles the waves are trying to pass through. And so it is with x-rays and atoms. The spacing between two iron atoms in a piece of steel is about 0.3 nanometer—right in the middle of the range of wavelengths for x-rays. Visible light, its wavelengths about a thousand times longer than x-rays, is reflected right off a shiny piece of steel. By contrast, the tiny wavelengths of x-rays allow a percentage of them to pass right through the similar-sized gaps between the atoms at the surface of a piece

---

enough of the light to make transmission over long distances impractical. Improved, high-purity glass was developed, and eventually fibers were produced that could transmit signals sixty-two miles or more.

of steel. Inside the steel, individual x-rays will eventually strike atoms, and some of that radiation will then bounce back out of the steel. This is diffraction. The x-rays are "interacting with a pattern of obstacles," the obstacles being the atoms in the steel. The amount of radiation of a given wavelength that penetrates the steel and then re-emerges depends on the precise spacing and patterns among the atoms and the exact wavelength of the x-rays, among other things. The mathematics of diffraction can get pretty complicated. But it's worth the trouble, because x-ray diffraction is a powerful technique for identifying materials based on the patterns and spacing of their atoms. Take a pinch of table salt, scrape some rust off an old nail, add a little chalk dust, throw in a few grains of sand, and grind it all up with a mortar and pestle. If you run the resulting concoction through an x-ray diffraction machine, all the chemical compounds in the sample will be precisely identified, even for tiny fractions of a gram. X-ray diffraction is so precise because the patterns and spacing of atoms in a crystalline compound, such as table salt, are as unique as the fingerprints, or the DNA, of a human being.

## Light and Color

Moving from longer wavelengths to shorter, the visible-light spectrum passes from light that most of us perceive as red to orange, yellow, green, blue, and finally violet. The naming scheme for the colors in visible light makes sense. The longest visible wavelengths are red, residing right next to the slightly longer (but invisible) infra*red* waves. At the other end of the visible spectrum, violet light has wavelengths a little longer than those found in the adjacent ultra*violet* region.

When Isaac Newton first described the visible spectrum, he included indigo in between blue and violet (hence the familiar mnemonic ROY G BIV), but these days most authorities have eliminated indigo. Whether we refer to the visible spectrum as having six or seven colors is beside the point, really, since it is generally agreed that the human eye can distinguish, crammed in between red and violet, about ten million different colors. Why that is so is mostly saved for part 2, "Sensation," and part 3, "Perception." But the story of color begins here, in the world of stimulus, and it begins with my green shirt.

I used to have a lovely (to me) light green dress shirt. I would sometimes wear it with blue jeans or, for a dressier look, with a pair of darker slacks. As a result of several color misadventures early in my career, I've learned to look myself over carefully in the bathroom mirror every morning before I leave for work. And this particular outfit—light green dress shirt with, say, dark blue slacks—passed the test at home. There, my bathroom mirror is nearly always bathed in plenty of natural sunlight during the morning hours.

But when I arrived at work for the first time in this particular getup, I received a bit of a shock. The shirt that at home had appeared to be a very natural light green, about the color of a dried bay leaf, became something totally different under the fluorescent lights of the building I work in. There, and I'm sorry to have to put it so bluntly, my shirt appeared positively snot-colored. I can only imagine the messages my students were texting to their friends during my lecture that day: "Omg his shirt looks like snot lol," or something like that. I seem to have a short memory for incidents like this, and I wore the shirt to work several times before I finally remembered to weed it out of my closet and donate it to charity.

The colors we perceive are influenced by many things, but first and foremost they are influenced by the combinations of wavelengths in the light shining on the object we are looking at. Sunlight is different from incandescent light, which is different from fluorescent light, or halogen light, or candlelight. The light at noon differs from the light at dusk, and a sunny day offers different light than a cloudy one. Each light source produces its own spectrum of wavelengths, and the things we look at, such as green or greenish shirts, appear different depending on the light source.

If they want to stay in business, those who market shirts and other clothing have to realize how important lighting is to their bottom line. Paco

Underhill is the founder and CEO of Envirosell, a company that specializes, according to its Web site, "in the study of human behavior in retail, service, home, and online environments." Underhill was a pioneer in the application of anthropological research techniques to the study of shopping and has written several books on the subject. The lighting in retail stores, he says, is something we rarely notice, which means the store's lighting designers probably did a good job. When we notice the lighting in such a store, it's usually because it's too dim, too bright, or too harsh.

The number of variables in the design of a lighting scheme for a retail store is daunting. There is the choice among different types of lamps, such as fluorescent versus incandescent versus halogen. The intensity of the light is important, too. Older shoppers require brighter light, due to age-related changes to the eyes. Younger shoppers generally prefer more subdued lighting. It pays for a retailer to understand the demographics of her clientele. The colors in a retail store are also important. Dark colors absorb more light, while lighter ones can be too harsh. Certain colors, such as yellow, are troublesome. Underhill says that yellow used in advertising has fared poorly in some of his studies, particularly among older shoppers. Best to stick to black, white, and red. Another variable in retail lighting is the time of day, which changes the amount of natural lighting available, especially for street-level stores, although not so much in the mall.

The product, Underhill tells us, does not sell itself, sitting there on the shelf at the store. Lots of other factors are required; effective lighting is only one of them. Chances are, I would never have bought that green shirt if it had been bathed in fluorescent light on the rack in the retail store where I found it.

When light impinges upon something, such as my erstwhile green shirt, some of the wavelengths in the light will be absorbed by the material. A black shirt absorbs, and turns into heat, nearly all the wavelengths and thus most of the energy in sunlight, and so a black shirt is a lousy choice if you're outside on a really hot day. A white shirt reflects nearly all of the visible radiation that hits it. At home in my sunlit bathroom, my green shirt absorbed all of the radiation in sunlight except for those wavelengths that, when combined, resulted in that lovely bay-leaf green. So why did that same shirt look so much worse under fluorescent light?

There are lots of different kinds of fluorescent light bulbs, each producing its own spectrum of light. Modern fluorescent lights tend to produce a color spectrum more pleasing to the eye than older ones did. In the bad old days, fluorescent light tended to produce a rather "spiky" spectrum, compared to sunlight, incandescent light, or halogen light, all of which produce a smoother, more continuous spectrum, as shown for sunlight in figure 2. Most of the energy in a fluorescent light is concentrated in just a few very narrow wavelength bands. Typically, the fluorescent spectrum is noticeably lacking in red light, which centers on about 700 nm. Red is a warm color, and since fluorescent light usually lacks it, many of us perceive fluorescent light to be somewhat cold and harsh.

A green shirt thus acts like a filter. Light waves, from the sun or a fluorescent bulb, strike the shirt. The shirt then absorbs or filters out some of that radiation. The rest of the light reflects off the shirt and enters the eye. What is reflected off the shirt is determined by two things: the nature of the light hitting the shirt, and the light-absorbing properties of the shirt fabric. Some of the wavelengths in sunlight that bounce off the shirt must be missing in the fluorescent light. But there is a second possibility for why the fluorescent light made my shirt look so bad, and that is the *extra* wavelengths in fluorescent light that may not be present in sunlight or that are much less intense in sunlight.

In the strange case of my light green shirt, what my eyes perceived as ugly under fluorescent light could be caused by either *missing* or *extra* wavelengths in the fluorescent light, as compared to sunlight.

When it comes to color, sometimes it makes more sense to think of things in terms of *addition,* whereas other times *subtraction* seems more logical.

## Subtractive Color Theory

The theories of color mixing are often poorly described to young students, and in any case they aren't as simple as we may vaguely recollect from long-ago school lessons. In *The Physics and Chemistry of Color*, Kurt Nassau contends that "if one is not confused by color mixing, one does not really understand it!"

In elementary school art class, I learned that when you mix two colors, you get a third. Blue plus yellow paint, when mixed, yields green. Red plus yellow gives orange. This is *subtractive* color mixing (even though it involves combining two paints). Each of the individual paints we mix subtracts (absorbs) a different portion of the visible spectrum, and thus the mixture yields a color different from either of the colors being mixed.

In subtractive color theory, there are three primary colors. These are cyan (bluish green), magenta (purplish red), and yellow.[3] When cyan, magenta, and yellow are mixed, each should, in theory, subtract out one-third of all visible light. Three times one-third equals one, and thus C, M, and Y, when mixed, should yield black. In the real world, this is just an approximation, and a CMY mixture typically yields a muddy dark brown. In practice, subtractive color theory often goes by the abbreviation CMYK. The K is for black. It's called "key." B isn't used for black, to avoid confusing it with blue. K, or black, is added to CMY in printing applications for various reasons. For one thing, black gets used a lot, especially for text, and it's cheaper to use black pigment than to have to add C to M to Y all the time. Also, C plus M plus Y produces black in theory, but in practice the resulting dark brown is found wanting.

In the world of color printing, CMYK is important. With those four colors, you can print anything. Humans may be able to distinguish ten million different colors, but with CMYK, and a computer to help with the mixing, you can match any of them.

Modern paint-matching technology is a practical application of subtractive color theory. Not so long ago, I would go to the paint store to try to match the color of, say, a paint chip from my dining room wall. The store had these huge collections of color samples, printed on little cards,

---

[3] The three primary subtractive colors are often presented as blue, red, and yellow, but this is incorrect.

to which I could painstakingly compare my paint chip, using the only instrument available: my eyeballs. Lousy matches were commonplace, at least for me.

These days, the stores still have collections of color sample cards, but the eyeball method of color matching has largely been replaced by the spectrophotometer, an instrument that measures light intensity across the visible spectrum. The machine bounces a beam of white light off your color sample.[4] The color sample absorbs (subtracts) certain wavelengths in the light and reflects the rest into the instrument's detector. There, the light is passed through a series of filters, one at a time. The first filter might remove all the light except that in the 400–410 nm range. The second filter might remove all the light except the 410–420 nm wavelengths, and so on, up to 700 nm. A photocell behind the filters measures the intensity of light that passes through each filter. The more filters there are, the more precisely the light reflected off the color samples can be measured.

Once the instrument has finished measuring the light that passes through each of the filters, it uses software to convert the results into a recipe for the paint you need to match the color sample. You may have seen the results on the adhesive label they attach to the can of matched paint. So many ounces of this color and that color, added to basic white paint, yields a color that matches your color sample pretty well, most of the time.

"Most of the time" seems to mean about 90 percent of the time. As we saw with the CMYK story, subtractive color theory does not always produce results pleasing to the eye. Human color vision is extremely refined. A spectrophotometer, as described above, might chop visible light up into 10 nm chunks, but the human eye can discern color changes far smaller than that. So it's not surprising that the recipe the machine spits out results in an unacceptable match about 10 percent of the time. What to do? At most stores they take a sample of the newly matched and mixed paint, brush it on a piece of paper, and dry it with a blow dryer. If it doesn't

---

[4] If the white light in the spectrophotometer is very different from the light in your home (recall my green shirt), the match you get may not be very good.

match, that is, if it doesn't please the customer's eye, the technician at the store can go freestyle. Using her practiced eye, she can add a few drops of this or that color to obtain a better match than that yielded by the spectro-photometer. These instruments work pretty well, though, and they've become so inexpensive that decent-quality models are affordable for many do-it-yourselfers.

That's subtractive color theory in a dark brown nutshell. What about addition?

## *Newton and Additive Color*

Perhaps the single most important event in the history of the science of optics occurred early in the year 1666, when Isaac Newton passed a beam of sunlight through a small hole into his otherwise darkened bedroom. He placed a triangular glass prism in front of the sunbeam. The prism refracted (or bent) the rays of light, resulting in a rainbow-like display of colors on the opposite wall. Newton termed this a "spectrum" of color, and he observed that the colors produced by his prism varied from red to orange to yellow to green to blue to indigo to violet (ROY G BIV, although, as noted earlier, indigo is generally left out these days). Newton based his rather arbitrary choice of seven colors by analogy to the seven notes (A–G) of the musical scale.[5] But Newton also understood that the color in light is simply what he, or any person with normal vision, would perceive. Light isn't "colored" any more than odors are sweet or sour, or vibrating air molecules are harmonious or dissonant. These are all things we perceive.

In any event, Newton did not stop with his first prism. He employed a second prism to gain even deeper insight into the nature of light. The second prism was turned upside down in relation to the first, as shown in figure 3. The color spectrum emitted by the first prism passed through the second prism, where it was refracted again. And what came out the

[5]More on the connections between music and color in part 3.

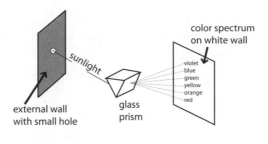

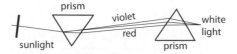

Fig. 3. Newton's prism experiments.

far side of that second prism was a beam of white light, essentially indistinguishable from the sunbeam that entered the first prism. This was the birth of the additive theory of color.

When the color spectrum emitted by the first prism passes through the second prism, the rays of individual colors, red through violet, are *added* back together. Their sum is just the spectrum of visible light that is produced by sunlight.

When light *rays* are combined, their colors *add* to each other. This stands in stark contrast to the *subtractive* behavior observed when mixing several different colors of paint, as described earlier. In addition to Newton's prisms, various other classic and easily reproduced experiments prove the additive theory of color. One such experiment involves three light projectors (strong flashlights will do) and three colored filters—one red, one green, and one blue. When the three lights are projected on a white wall, one observes, not surprisingly, a red circle, a green circle, and a blue circle. When the red circle partially overlaps the blue circle, the region of intersection appears yellow. This much is not especially surprising. But when all three circles are partially overlapped, the area of intersection at the center where all three beams are shining is white.

Adding the red, green, and blue beams together to get white light does essentially the same thing as Newton's second prism did to his color spectrum. When all the fundamental wavelengths are added back together, the full, undifferentiated spectrum is obtained, which we perceive as white light.

When we add red to green to blue (RGB) light waves, we get *white* light. But when we add cyan to magenta to yellow paint (CMY), we get *black* paint. You may be starting to understand why Kurt Nassau says that if you're not confused by color theory you don't really understand it. Light waves *add* colors when combined. Pigmented objects are things we look at, *things off of which light reflects.* These *subtract* from the color spectrum when combined. If we add enough light waves to one another (red plus green plus blue yields the whole visible spectrum), we get the color that represents "all light," and that color is white. And if we add enough *pigments* to one another (cyan plus magenta plus yellow covers the entire spectrum), we get the color that represents "no light," and that's black.

The additive world of RGB goes head-to-head with the subtractive world of CMY every day, in the monitor of your personal computer and its printer. Your computer monitor is an RGB device, whereas the printer comes from the world of CMYK. This creates challenges for computer hardware and printer manufacturers, since we've all gotten so used to being able to print out exactly what we see on our computer screen. What you see is what you get (WYSIWYG), or at least we hope so. The computer monitor is an RGB additive device in that it produces various colors by adding red, green, and blue light together in the right proportions and at a fine enough level to produce the colored image desired. How this works depends on the technology of the monitor.

Most computer monitors these days are LCD, or liquid crystal displays. They work like this: white light shines from the back of the screen to the front. In a layer sandwiched between the back and the front of the screen is a network of extremely fine "liquid crystals." When one of these tiny crystals receives a jolt of electricity, it changes its crystalline form and becomes opaque, thus blocking light for that little region of the screen. On top of these tiny crystals are positioned equally tiny red, green, or blue filters. By turning on or off the light-blocking ability of each tiny crystal, the screen controls how much red, green, or blue light gets transmitted and in what pattern.

When you ask the printer to produce a paper copy of the image on your screen, it is necessary to transform the RGB light image on the screen into the CMYK terms that the printer will use. Some printing technologies use variations on CMYK that go by other acronyms, and often they use a few more colors, but the basic idea is the same: to synthesize millions of individual colors from a very small number of primary colors. My CMYK printer thus has four ink cartridges: cyan, magenta, yellow, and black.

Most of the printed images we look at every day are obtained by laying down tiny dots of color in patterns fine enough that they appear to be continuous. Printing things with patterns of dots like this is sometimes called halftone printing or screening. It's relatively easy to see the individual dots on newsprint, especially with the aid of a magnifying glass. A color weather map in the newspaper is a good place to look for dot patterns. With higher-quality printing, the dots are more difficult to see without the aid of a microscope. In CMYK printing, the dots of each individual color are laid down in straight, parallel rows, but the rows of each color are laid at a different angle to the other colors. This is to avoid interference patterns among the different colored dots.

Printing accomplished in this manner is reminiscent of the pointillist painting techniques of Georges Seurat and other postimpressionist painters. Seurat's masterpiece *A Sunday Afternoon on the Island of La Grande Jatte—1884* is perhaps the best-known example of this technique. In pointillist painting, as in halftone printing, we perceive the colors of the pure, individual dots as blended into other colors through a process sometimes called optical blending. Seurat believed that painting in small dots and thus not physically blending his pigments would make the colors more vivid. It has been estimated that there are about three and a half million individual dots of color in Seurat's *A Sunday Afternoon,* a painting that took two years to complete. That works out to about forty-eight hundred dots per day. The technique seemed to work for him, and it certainly works in the world of color printing, where computers make the whole process somewhat quicker.

# Chapter 2 • Chemical Stimuli

## Smelling Airborne Molecules

What we smell are molecules in the air. That much is clear. Early in my career, I worked as an engineer in the petrochemical industry. I remember my first trip to a plant where my company produced ammonia. Entering the plant for the first time, I was hit by a blast of ammonia odor so strong it nearly knocked me over. For a moment, I couldn't even speak. Finally, I asked my colleague, who had spent years in the plant, how he could possibly stand to work in the presence of such an overwhelming odor. "What odor?" he replied with a grin. With something approaching horror, I realized he wasn't kidding.

Ammonia is not a classical odorant, since it is not sensed by the same receptors in the nose as most of the other things we smell. With chemicals such as ammonia, certain other noxious substances, and even things like menthol, the stimulus is not received by the nasal receptors, but by something called the common chemical sense. Thus, strictly speaking, we don't "smell" ammonia. Someone who has lost his sense of smell is thus likely to still be able to detect the presence of ammonia in the air. We'll return to this story in part 2.

Later, I reflected that what years in that plant had done to my colleague's ability to sense ammonia was probably analogous to what looking at the sun through his telescope eventually did to Galileo's eyes, or what thrashing out power chords night after night at over 120 decibels had done to Pete Townsend's ears.

Perhaps. But we know why Pete's thunderous chords sound the way they do, and we know why the sun looks the way it does. We don't know, in nearly so much detail or with certainty, why any given odorant smells the way it does. We can't even predict whether a given airborne molecule will have any sort of smell at all.

Just what is the stimulus here? The Greek philosopher Democritus (460–360 B.C.) speculated that what we smell are different "atoms" of various sizes and shapes that come from objects. Other philosophers, Aristotle among them, came up with all sorts of crazy ideas about where smell comes from. It wasn't until the 1700s, with the kinetic theory of gases, that Democritus's speculation was confirmed.

Smell, considered the oldest, genetically speaking, of the senses, remains in some ways the most mysterious. And much of the mystery relates to the stimuli involved. At the simplest level, we do understand the stimuli of odor. Our olfactory receptors are stimulated by airborne molecules. When someone on the other side of the room lights a cigar, it takes time for your nose to get the signal. The time it takes before you sense cigar smoke is the time it takes for the molecules in the smoke to be wafted along on the relatively feeble air currents of the room and then eventually to make their way inside your nose and onto the smell receptors there. Whereas vision proceeds at the speed of light and hearing at the speed of sound, olfaction works at the comparatively glacial pace of the air currents that carry odorant molecules from point A to point B, inside your nose.[1] So, that's the stimulus:

---

[1] It is sometimes assumed that what conveys odorants to the nose, especially indoors, is the process of molecular diffusion. This is almost never the case. Molecular diffusion, even in the gas phase, is far too slow. Even indoors, air currents from HVAC systems, from drafts, and from thermal gradients move odorants around much, much faster than diffusion alone.

The speed of light is nearly 300 million meters per second. The speed of sound in air is 343 meters per second. The air currents carrying cigar smoke in an indoor room are likely to be moving at well under 1 meter per second.

cigar smoke molecules. But what property or characteristic of those molecules do we sense?

The two main theories on what stimulates the sense of smell are sometimes given one-word names: shape and vibration. These refer, respectively, to the molecular shape and to the vibration patterns of the airborne molecules involved. In shape theory, it is the physical shape of a molecule that determines its odor. The atoms of a molecule bond to one another at precise angles. $H_2O$, water, is in reality H-O-H, where the angle from one H atom through the O atom at the vertex and on to the other H is 104.45 degrees. Thus, a water molecule has sort of a boomerang shape.

The more atoms a molecule has, the more complex its shape generally becomes. In shape theory, the unique shape of a molecule acts like a key that fits into a unique lock inside one of the smell receptors in the nose. And once the key is inside the lock, this theory maintains, its odor can be sensed by the apparatus of the nose.

There are several problems with the shape, or lock-and-key, theory of odor. One is that it requires a unique "lock" (receptor) in the nose for each unique "key" (odorant) that humans can perceive. Since it is generally accepted that a person can distinguish up to about 10,000 different odors, that would require a lot of unique odor receptors in the nose. The number of unique smell receptors is believed to be far less than this—perhaps around 350. There aren't enough locks for all those keys. Another problem with the lock-and-key theory is that some molecules with similar physical shapes have wildly different odors.

The vibration theory of odor holds that the nose operates something like a vibrational spectrometer, analyzing the vibrations of molecules and passing those on to the brain through the instrumentation of the nose.

Above the temperature of absolute zero, all atoms are constantly vibrating. When atoms are combined together into molecules, the various ways in which molecules can vibrate gets very complicated. Water, with three atoms, can vibrate in three different ways or modes. Ethanol ($C_2H_5OH$) has nine atoms and twenty-one different vibrational modes. Bigger molecules have even more modes.

For a simple molecule like water, it's relatively easy to visualize each of its vibration modes. Water is a boomerang-shaped molecule, with hydrogen atoms on the two ends of the boomerang and the larger oxygen atom right in the middle. Imagine that each hydrogen atom is connected to the oxygen atom by a stiff spring. If you stretch both springs the same amount at the same time, and let go, the molecule will vibrate in the "symmetric stretching" mode. If you bend the two springs toward each other and let go, you get "bending" mode. The third mode, "asymmetric stretching," occurs when you stretch one spring more than the other.

Various types of instruments have been developed for measuring the vibrations of molecules. These can be used to identify molecules, much as fingerprints or DNA identify people. The frequencies of molecular vibration are often in the infrared region of the electromagnetic spectrum, and these can be measured using an infrared spectrometer, which detects IR waves.

The vibration theory of odor holds that the nose is an instrument analogous to an infrared spectrometer. This means that the nose identifies molecules based on their unique vibrational modes. The identifications are then passed on to the brain. In the case of a gas like hydrogen sulfide ($H_2S$), what we perceive is a smell like rotten eggs.

How can "vibration" become "rotten eggs?" Vibration is, well, so math-y. The smell of rotten eggs is so visceral, the seeming antithesis of math. The sense of smell is so ancient, genetically speaking, that it seems counterintuitive that we could describe it with math. But color can be pretty visceral too, and we have come to terms with its precise mathematical description. Why not odor as well?

Part of the problem is that the vibration theory of odor is by no means universally accepted or understood. This is not for lack of effort on the part of Luca Turin, perhaps the best-known and most eloquent proponent of the vibration theory. Turin moves effortlessly between the cold, scientific world of molecular vibrations and the vivid, visceral, and utterly nonquantitative world of smells. His many publications include an article in the journal *Chemical Senses* entitled "A Spectroscopic Mechanism for Primary Olfactory Reception" and a book called *Perfumes: The Guide* (co-authored with Tania Sanchez). Turin's career is chronicled by Chandler Burr in the delightful book *The Emperor of Scent: A True Story of Perfume and Obsession.*

By all accounts, Turin has a nose whose sensitivity would give a bloodhound a run for its money and an analytical mind that has led some to believe he may be worthy of the Nobel Prize for chemistry. *Perfumes: The Guide* can be read for entertainment as well as instruction. Witness the description of *Hugo*, the men's cologne from Hugo Boss: "Dull but competent lavender-oakmoss thing, suggestive of a day filled with strategy meetings."

One consequence of the lack of a comprehensive theory of smell is that it leaves those charged with creating new odors somewhat in the dark, if I may be allowed to mix the senses in this way. Perhaps you were unaware that these people exist, but they do, and there's big money in their game. Their work is analogous to the pharmaceutical industry, wherein complex new molecules are synthesized for their abilities to control or prevent diseases. The stakes are admittedly somewhat lower in the odor game, as lives do not hang in the balance, but nonetheless, millions of dollars can be lost in the development of a new odor that doesn't perform as intended, or gained if it does.

The odors of the molecules of a laundry detergent, the molecules responsible for cleaning your clothes, are by themselves unlikely to induce the particular feelings in the consumer that are desired by the manufacturer. Those feelings might be described with words like "clean, fresh, and rosy." If the cleaning molecules themselves can't do it, the manufacturer is obliged to mix in some other molecules that *do* evoke those sorts of feelings, since the odor of a product like laundry detergent is considered an important marketing tool. The next time you walk down the detergent aisle of your supermarket, pay attention to your nose. That particular aisle smells like no other in the store.

The synthesis of new molecules developed for their particular odors is an important, if somewhat obscure, business. It's also poorly understood scientifically. New molecules can be synthesized so that every detail is known about their structure, but practically nothing can be predicted as to their odor. Imagine the same situation relative to color or sound. What if we couldn't predict the additive or subtractive results of mixing colored lights or pigments? I simply can't imagine this. "Every now and then a man's mind is stretched by a new idea or sensation, and never shrinks back to its former dimensions," said Oliver Wendell Holmes. It's hard to imagine the world without color theory or its benefits, such as my color computer monitor and color printer. We can't say the same thing about odor.

The problem of not understanding the mechanism of smell is compounded by a language problem. The stimuli of vision can be described in terms of the frequency and intensity of the light waves that reach the eye. On the other side of the eye, frequency and intensity are perceived as color and brightness. We have language for the stimulus and language for the sensation. The same is true for hearing, but the sense of smell is an exception. We use the same language (odor, scent, smell) to describe both the stimuli and the sensations of olfaction.

## Taste Stimuli

What we smell are molecules in the air, and what we taste are molecules dissolved in water. The machinery of the lips, teeth, salivary glands, and tongue serve to prepare these watery recipes and present them to the sensory receptors, called taste buds.

While smell is a gas-phase phenomenon, taste takes place in the liquid phase. Perhaps as a result, the things we can taste are relatively limited, compared to the ten thousand or so different odors we can differentiate. Here, we refer to differences in kind (sweet as opposed to salty) and not degree (mildly versus sickeningly sweet).

Sweet, salty, sour, and bitter tastes are well known. Umami, or savory, has been acknowledged for some time as a fifth taste category. Others are said to exist, depending on which expert you consult.

Sweet tastes come from naturally occurring sugars and from a growing number of powerful artificial sweeteners. An artificial sweetener such as neotame can be ten thousand times as sweet as natural sugar. As a result, five or six tiny grains of artificial sweetener can replace about ten teaspoonfuls of natural sugar in one twelve-ounce soda. Since they are added to foods in such tiny amounts, it's no wonder that artificial sweeteners contribute so little to the caloric value of a food or beverage. Whether they are detrimental to your health, especially in the long term, is another question. It's also worth pointing out that as the quantity of artificial sweeteners in the average American diet has increased, the amount of sugar we consume has gone up as well. We'll look closer at those numbers, and at artificial sweeteners, in part 2.

# Other Chemical Senses

Smell and taste are the best-known chemical senses, but they are not the only ones. What has come to be known as the common chemical sense is an important part of our ability to sense the various things that enter our noses and mouths. Why does a spicy food, such as a jalapeño pepper, cause such a hot sensation when we ingest it? Hold a jalapeño up to your nose and smell it. There is no hot sensation. The hot sensation is likewise not detected by any of the various types of taste buds on the tongue. But that hot sensation is nonetheless real. The hotness of peppers, the coolness of menthol, and the textures of foods are all important stimuli. But none of these are detected by either the smell receptors in the nose or the taste buds in the mouth.

The hotness of peppers is measured on the Scoville scale, figure 4. Wilbur Scoville was an American chemist. He published his technique, an innovative blend of analytical chemistry and good old-fashioned taste

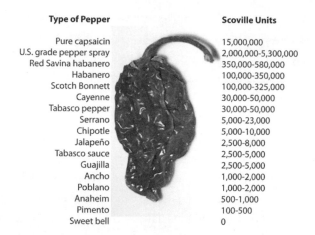

| Type of Pepper | Scoville Units |
| --- | --- |
| Pure capsaicin | 15,000,000 |
| U.S. grade pepper spray | 2,000,000-5,300,000 |
| Red Savina habanero | 350,000-580,000 |
| Habanero | 100,000-350,000 |
| Scotch Bonnett | 100,000-325,000 |
| Cayenne | 30,000-50,000 |
| Tabasco pepper | 30,000-50,000 |
| Serrano | 5,000-23,000 |
| Chipotle | 5,000-10,000 |
| Jalapeño | 2,500-8,000 |
| Tabasco sauce | 2,500-5,000 |
| Guajilla | 2,500-5,000 |
| Ancho | 1,000-2,000 |
| Poblano | 1,000-2,000 |
| Anaheim | 500-1,000 |
| Pimento | 100-500 |
| Sweet bell | 0 |

*Fig. 4.* The Scoville scale for the hotness of peppers, shown with a photo of a dried habanero pepper. (Data from eatmorechiles.com; photograph by the author)

testing, in 1912. The chemical that renders peppers hot is called capsaicin. Scoville used alcohol to extract capsaicin from samples of dried pepper. The alcohol extract is diluted in a sugar water solution. The resulting concoction is fed to taste testers. The weakest solution (the least amount of capsaicin per volume of sugar water) for which the taste testers can detect a hot sensation is found in this manner. From this, the Scoville rating is determined. For example, the extract from a certain type of jalapeño pepper might have to be diluted five thousand times to reach the threshold of detectability. The Scoville rating of that pepper would thus be 5,000.

With Scoville ratings from 2,500 to 8,000, jalapeño peppers are relatively mild. Habanero peppers, available in most U.S. grocery stores, are rated at 100,000 to 350,000 Scoville units. Recipes with habaneros often recommend that the chef wear rubber gloves when dealing with these innocent-looking little fellows and that all cooking implements be carefully cleaned afterward, to avoid any possibility of getting habanero oils in the eyes or on other sensitive areas of the body.

The hottest peppers are even hotter than habaneros. Some exotic varieties are rated at up to 1 million Scoville units. Pepper sprays available in the United States might rate anywhere from 2 million to over 5 million Scovilles. Pure capsaicin tips the scale at 15 million.

It is possible to measure the hotness of peppers without taste testers, using the purely chemical technique of liquid chromatography. The Scoville scale remains popular, however. I've seen it posted in several grocery stores. There's even a scale to convert hotness measured by liquid chromatography to Scoville units.

# Chapter 3 • Mechanical Stimuli

## Sound Waves

A few years ago, my family and I moved into a new home. Being real estate veterans, my wife and I looked the place over quite carefully before taking the plunge and buying it. And we're generally pretty happy there. It's a lovely place on a leafy, out-of-the way street without much traffic. Nice and quiet—just what we'd been looking for.

I was somewhat surprised, then, on one of the first mornings after we moved in, at what I heard when I went outside to collect the newspaper at about six o'clock. In the neighborhood, not a creature was stirring, but the traffic noise from the interstate highway, about eight-tenths of a mile away, was clearly audible. Not nearly so loud that I would need to raise my voice to be heard, but audible nonetheless. The dull whine of tires spinning down the road, carrying vehicles at upwards of seventy miles per hour, is the most persistent characteristic of this noise, punctuated at regular intervals with the throaty percussion of the diesel engines of the big rigs, especially when their drivers back off the throttle.

This was more than a little annoying. I was particularly miffed that I hadn't noticed this noise on any of the several long visits we'd made to the place while house hunting. I was also shocked that eight-tenths of a mile wasn't far enough away to avoid this problem. Since then, I've noticed that the highway noise at that time of day is nearly as loud at up to a mile and a quarter away from the interstate, based on data carefully gathered while walking the dog.

Happily, the noise fades quickly as the day progresses, and by midmorning it is generally imperceptible. To add to the mystery, there are some days when I can't hear the traffic at all, even at six in the morning. What's going on? It's certainly not the traffic patterns on the interstate. There are plenty of cars and trucks at all hours of the day and night.

This problem can be explained by a well-known meteorological phenomenon combined with the physics of sound waves. Early in the morning, a layer of colder air is sometimes trapped close to earth, the sun having heated the air above it more quickly. The boundary between the cool air and the warmer air above it can act as a barrier to reflect rising sound waves back toward the earth, and toward my house, in the case of our highway noise. The greater the temperature difference and the sharper the gradient between cool and warm air, the more likely this is to happen. In order for this reflection phenomenon to occur, the sound waves have to strike the boundary between the warm and cool layers of air at a very shallow angle. Imagine skipping a flat stone across a still lake. For the stone to skim across the lake, it has to be traveling nearly parallel to the surface of the water. If your throw angles the stone downward too sharply, the stone won't skip but instead plunges immediately under the surface of the water, never to be seen again. Likewise, the only sound waves that can be reflected off the boundary between air layers, and then back to earth, are those at a shallow angle to the boundary. Later in the day, when there is little or no temperature difference between the layers of air, the noise from the freeway simply travels up into the atmosphere.

Sound waves, the stimuli of hearing, are the province of acoustics, the science of sound. Something serves as the source: a plucked guitar string, a baseball striking a bat, hot gases being expelled from a jet engine, for example. The sound source causes the air around it to vibrate. These vibrations travel away from the sound source, just as water waves travel outward from a stone dropped in a still pond. When the sound waves reach your ear, the sensation process begins. We call it hearing.

Light and other electromagnetic waves can travel through a vacuum, such as outer space. Not so with sound waves. They require a medium—some sort of physical substance through which the waves are transmitted. Often, that medium is air. But sound waves are perfectly at home in other media, such as water or other liquids, or solids such as wood, metal, or bone. The sounds we hear arrive in our inner ear through two different media: the air, which vibrates our eardrums, and the bones in our skull, which transmit vibrations directly to the inner ear, bypassing the eardrum. This is perhaps the main reason why we are often startled by the sound of own voice when we hear it played back in a recording. We're used to hearing our own voice as we create its sounds, beginning in the larynx and moving out through the mouth. Much of what we hear when we speak arrives at our inner ears through bone conduction. But when we hear our recorded voice through a loudspeaker, almost everything we hear is airborne sound. Subtracting out some of the bone-conducted nuances in what we hear in our voices makes a big difference. That difference is striking, and it's no wonder we sometimes don't even recognize our recorded voice, especially the first time we hear it.

When waves of any kind travel through a medium, the waves move but the medium itself doesn't. As waves roll gently past a fishing boat in the ocean, the boat moves up and down, vertically, but the waves don't displace the boat horizontally. Likewise, the medium that sound waves travel through doesn't travel with the waves. This seems intuitive for sound traveling through solids, but perhaps less so for air.

The wind can influence the motion of sound waves. If the wind is blowing strongly out of the south, for example, it is more difficult for someone south of you to hear you yelling at her than someone north of you the same distance away. The wind does *not* blow the sound waves back in your face—that's not how sound waves work. But wind speed generally does increase with altitude, as any pilot or wind-turbine expert would tell you.

As the sound waves move away from your mouth, the waves that are angled slightly up are traveling in slightly faster-moving air. The speed of the sound waves remains the same relative to the air, even though air is moving (the wind is blowing). Relative to the stationary ground, however, the sound waves are moving faster as their altitude increases.

When waves of any kind move from one medium to another in which they travel faster, they tend to bend and change direction. A familiar example is visible-light waves traveling from air into water. Light travels slower in water than in air, and thus light waves are bent when they pass from air into water. Hold a soda straw at an angle in a clear glass full of water, and the straw appears to be bent.

Sound waves do the same thing when they pass from one medium to another. So when sound waves move at an angle upward from slower wind to faster wind, they are in essence gradually speeding up. The sound waves will thus bend upward slightly, so that some of them fail to reach their target, the person upwind of the one doing the yelling.

Waves of any kind moving through any medium have various properties. One of them is their frequency. The technical meaning of frequency is very similar to its general meaning, which is "how often." How often a wave passes a particular point is its frequency. Drop a series of stones into a still pond, and ripples (waves) of water will radiate outward from the point of impact. The rate at which the crests of the waves pass a certain point is the frequency of the waves—so many waves per second.

It's the same thing with sound traveling through air. The most common tuning fork produces a tone with a frequency of 440 hertz (Hz), or 440 cycles or waves per second, since one Hz is one cycle per second. A 440 Hz tone is the musical note of A above middle C and also goes by the names A440 and Concert A. If you are listening to the tone from an A440 tuning fork, sound waves are arriving at your eardrums at 440 Hz, or 440 ripples of the air per second.

Frequency is at the heart of the familiar Doppler effect. When an ambulance passes you, the pitch of its siren shifts. The pitch of a musical tone is what you hear. Pitch is closely related to the measured frequency of a sound. As the frequency of a sound increases, what we perceive—the pitch—also increases.

As the ambulance is approaching, you hear a higher pitch, which shifts to a lower pitch immediately as the ambulance passes and moves away. The Doppler effect was first described in 1842 by the Austrian physicist Christian Johann Doppler (1803–1853). He used the effect that now bears his name not to describe sound, but to explain the color of the stars based on their velocities relative to an observer on the earth. Sound waves change *pitch* based on the relative velocities of the sound source and the observer. Light waves, in analogous fashion, change *color*.[1] Doppler's work related to light waves, but others quickly extended it to sound waves. The Dutch chemist and meteorologist C. H. D. Buys-Ballot organized a demonstration in 1845 in which a trainload of trumpeters, all playing the same note, were transported past a group of stationary observers. The pitch dropped, as predicted.

In the Doppler effect, the apparent frequency is the frequency at which the observer meets the waves traveling outward from their moving source. The apparent frequency can be calculated from the actual frequency of the source, using the speed at which the waves are traveling (such as the speed of sound in air), and the speed at which the sound source is approaching the observer.

Imagine you are standing beside the road, and a car with an A440 tuning fork on the roof is approaching at 70 miles per hour. The speed of sound in air is about 761 miles per hour. The Doppler effect tells us that as the tuning fork is *approaching* at 70 miles per hour, the 440 Hz tone shifts upward to 483 Hz. As the tuning fork passes and recedes into the distance at 70 miles per hour, the apparent frequency shifts down to 403 Hz.

---

[1] A colleague of mine, a physics professor, has a red bumper sticker on his car that reads, "If this sticker is blue, you're driving too fast." A little physics humor, courtesy of the Doppler effect.

The frequency shifts in this example are more than a musical semitone, or half tone, in each direction. A semitone is the difference between two adjacent keys on a piano. Doppler shifts much, much smaller than this are clearly detectable by the human ear.

The Doppler effect has lots of applications in human-made instruments, including everyone's favorite, the policeman's radar gun. The radar gun uses the Doppler effect not on sound waves but on microwaves. These have much higher frequencies and velocities than sound waves. A radar gun might operate at 10 gigahertz, or 10 billion cycles per second. The radar gun, often handheld, directs a beam of microwaves at the target. Those microwaves travel at the speed of light, about 300 million meters per second. But the car is traveling far slower, say, 70 miles per hour, or 31.3 meters per second. In other words, the microwaves are traveling nearly 10 million times as fast as the car. So the microwaves bounce off the speeding car and are reflected back to the radar gun, which captures the reflected waves and measures their frequency. The Doppler effect uses the change in frequency from the transmitted to the reflected microwaves to calculate the speed of the target. Because the speed of the car is so small compared to the speed of light, the frequency change is also tiny—it might be only one part in 5 million. Nonetheless, such frequency changes are accurately measured and have been for many years. The first police use of this type of technology against speeding cars in the United States was in 1954.

Radar guns are not without their problems, not the least of which, from the standpoint of the police, is the prevalence of high-quality, low-cost detection units available to drivers.[2] A new technology has been developed that renders generations of such detectors obsolete. The new technology is called lidar. *Lidar* means "light detection and ranging," just as *radar* means "radio detection and ranging." Lidar guns, similar in appearance to radar guns, send pulses of infrared radiation at their targets. But lidar does

---

[2]I've always wondered why a device whose sole purpose is to facilitate lawbreaking could possibly be legal. But not only are these devices legal most places in the United States, they are the foundation of a billion-dollar industry.

not use the Doppler principle. Instead, it simply measures the time it takes for one IR pulse to travel from the gun to the target vehicle and back again. Since the speed of the infrared signal is known, the system can calculate how far away the target is. By sending out hundreds of pulses per second, the lidar system effectively tracks the motion of the target. It knows the position of the target as a function of time, which is just another way of saying that it knows the target's velocity.

## Stimuli of Touch

At the clothing store, when we want to feel a piece of fabric such as a silk shirt or a cashmere sweater, we do so by rubbing our fingers across the surface. Why isn't simply touching the fabric good enough to provide the stimuli we need? It turns out that we are very poor at sensing the difference in surfaces with fine textures based on touching alone. "Fine" in this case means a surface finish corresponding to sandpaper somewhere between 120 and 150 grit. The average person can distinguish, through touch alone and without moving the fingers across the surface, the difference between 40 and 60 grit sandpaper. But the same is not true for 240 and 320 grit sandpaper. To distinguish those surfaces requires rubbing the fingers across the surface. Scientists refer to a "duplex theory" of texture perception. For rough surfaces, your fingers discern, without moving, how far apart the bumps are. For finer surfaces the fingers must move across or rub the surface.

Surface roughness is hardly the only quality we can discern with our fingertips. We can feel all kinds of other things about a surface, such as its hardness, wetness, and greasiness, its temperature, and whether it is vibrating. Not all of these are touch stimuli. That is, not everything we sense about a surface is stimulating the touch receptors in our fingers. There are lots of other receptors in there, such as those for temperature, body position, and pain, and they work in concert with the touch receptors.

The stimuli for the sense of touch all relate to deformation. When you rub your finger across a surface, the skin is deformed by the various bumps on the surface. Likewise, when you bend a hair on the back of your hand, that deformation is sensed, at the root of the hair shaft, below the surface of the skin. Cats, rats, and other creatures are quite sensitive to deformations of their whiskers.

I love to cook, and I'm particularly fond of my charcoal grill. One challenge when grilling is knowing when a piece of meat is ready. There are various tricks for figuring out when a steak on the grill is rare, medium, or well done. If you press on a grilling steak with your fingertip, its relative hardness is an indicator of doneness. Many guides to this technique include rather annoying qualitative standards such as, "medium rare—yields gently to the touch," and "medium—yields only slightly to the touch." The novice chef can't do much with that sort of guidance.

But there is a more sophisticated version that works pretty well. You can compare the hardness of a grilling steak with the hardness of the fleshy part of your thumb on the palm of your hand. First, touch your thumb to your index finger, making an "OK" sign. Now use your other index finger to touch the fleshy part of the thumb you're making the OK sign with. The relative softness you sense is about the same as a rare steak. Now, in your OK sign, replace your index finger with your middle finger. This stiffens the muscles at the base of the thumb. The flesh is less soft—about the same as a medium-rare steak. Now replace your middle finger with your ring finger. The base of the thumb stiffens a bit more—about like a medium steak. When you move on to your pinkie finger, you get a medium-well steak. This technique works pretty well for me.

In terms of stimuli, probing a relatively soft surface like a steak or your thumb involves more than the sense of touch. The motion of the fingertip as it moves five or ten millimeters or more into a soft surface also requires proprioception, or position sensing.

# Acceleration

## *Position*

Here's a nightmare scenario: You've been invited to join your boss for a home-cooked dinner. While her husband prepares the meal in the kitchen, you and the boss enjoy a cocktail on the veranda. When the meal is ready, you're called to the table. But as you cross the threshold into the formal dining room, cocktail glass in hand, you stub your toe on the edge of a thick Persian rug that must have cost at least twenty grand. Your head pitches forward, as does the nearly full glass in your hand. Time seems to stand still as you imagine the consequences to your career if you take a header and splatter your strawberry daiquiri all over that rug.

Miraculously, you regain your footing and avoid spilling a single drop of your colorful beverage. Cracking a lame joke about how clumsy you are, you take your seat at the table. Everyone laughs, breathing a sigh of relief.

It was your senses that saved you here, but which ones, and how? And what were the stimuli? The human body possesses the ability to sense the positions of its parts and the ways in which each part is moving. This sense is supremely precise in detecting the movement of the head. It's difficult to overstate the importance of this sense, and not just for dinner parties at the boss's house.

A few more quick examples: Close your eyes, raise your hand, and extend two fingers. How did you know it was two fingers, and not one or three? That's proprioception. Now stand in the middle of a room with no windows, and ask someone to turn off the lights. Did you fall down in the darkness? I'm glad you didn't; that's another example of proprioception.

Webster's New World College Dictionary defines proprioception as the "normal awareness of one's posture, movement, balance, and location based on the sensations received by the proprioceptors." We'll discuss the proprioceptors in part 2. For now, let's focus on the stimuli in proprioception. Light waves stimulate vision. Sound waves stimulate hearing. What stimulates our "awareness of posture, movement, balance, and location"?

Proprioception, as Oliver Sacks points out in *The Man Who Mistook His Wife for a Hat*, derives from the Latin *proprius,* or "one's own." Proprioception relates to being aware of or receptive to one's own self. The stimuli here are altogether different from anything we've discussed. These are not waves of any kind, nor are they related to the size, shape, or vibrational qualities of molecules.

What we're talking about, instead, are the position, velocity, and acceleration of the body and its various parts. This is the physics of motion, pure and simple.

## The Physics of Motion

It was Isaac Newton who, when he wasn't deriving the laws of optics or inventing calculus, determined the relationship between force and acceleration. The second of his famous laws of motion (Newton's Second Law) states that force equals mass times acceleration. Acceleration is just the rate at which something is changing speed. If you are traveling in a car on a straight road at a constant speed of sixty miles per hour, your acceleration is zero. If you slam on the brakes, the vehicle's speed drops rapidly. *Now* the car is accelerating, or decelerating in this case. And you can sense it.

As the car's speed decreases, your body's speed may or may not decrease along with it. That depends on whether you're wearing your seatbelt. Newton's First Law states that a body in motion tends to stay in motion unless acted on by some external force. In this case, let's hope the external force is supplied by your seatbelt. Otherwise it will be supplied, somewhat more crudely, by the dashboard of the car, if you're in the front seat.

I'm sure you remembered to fasten that seatbelt, and so we can determine the force it exerts on your body as the car slows down. That force, in pounds, can be calculated from Newton's Second Law. If you decelerate uniformly from sixty miles per hour to zero in 2.73 seconds, the average force exerted on your body by the seatbelt will be equal to your weight. So, if you weigh 150 pounds, a force of 150 pounds is pushing you forward horizontally into the seatbelt. An equivalent force, by the way, is always pushing you *downward* into the seat of the car, regardless of how much or how little the car is accelerating. The name for the force pushing you downward into your seat is your weight. Since the force pushing you downward is equal to the force of gravity, and that, in this carefully chosen example, just happens to be equal to the force pushing you horizontally into the seatbelt, we say the horizontal force is "one g," or one times the force caused by gravity.

Fighter pilots perform maneuvers during evasive actions in which they command their planes to undergo huge accelerations. As a result, their bodies are subjected to much larger forces than in our car example—perhaps five or six g's or even more. These acceleration-induced forces can even be large enough to cause a pilot to lose consciousness and, soon thereafter, her life. Documented fighter-plane disasters have occurred this way. The human body simply wasn't designed to withstand that much acceleration. Fighter pilots wear special pressurized flight suits and are trained in techniques designed to reduce the effects of excessive g-forces, and in particular to avoid blacking out when massive accelerations force blood out of their brains. But much lower accelerations can play havoc with the pilots of any sort of airplane. The crazy accelerations induced by flight can fool the senses into thinking the plane is doing one thing when actually it is doing something quite different. Pilots are thus trained to react to what an airplane's instruments are telling them, not to their own perceptions.

In an airplane, your body can undergo linear accelerations in six different directions: up and down, forward and backward, left and right. That's true in a car, too, but generally to a lesser extent, especially during normal driving. Pilots sometimes refer to the six linear acceleration directions in terms of the directions of the acceleration-induced forces on their eyeballs. When you slam on the brakes in a car, for example, that's "eyeballs out," since the deceleration would result in your eyeballs moving forward, out of their sockets, were they not restrained by the skull and associated muscles and connective tissue. When an airplane is pulling out of a steep dive, that's "eyeballs down." When an elevator traveling upward slows down as it approaches your floor, that's "eyeballs up."

All of these accelerations (up, down, in, out, left, right) are linear. They relate, as in the braking-car example, to a change of speed in a straight-line direction. Your body senses all of these, but it can also senses changes in another type of acceleration, angular acceleration. This is the change in the speed with which something is rotating. A dragster is sitting at the starting line waiting for the light to turn green. When the driver punches the throttle, an enormous torque or twisting input is applied to the rear wheels, and they begin rotating very rapidly. That's angular acceleration. It is accompanied, in this case, by a linear acceleration—the eyeballs-back variety, possibly in excess of five g's initially—of the vehicle as it hurtles down the quarter-mile track.

## *Accelerometers*

Instruments for measuring accelerations, just a few short years ago, were considered somewhat exotic. Now they are so common that many cell phones and video game controllers come equipped with them. They're called accelerometers. An accelerometer measures accelerations by applying Newton's Second Law. There are a great many ways to build a device like this, and the accelerometers on the market reflect this diversity.

In the example of a car braking from sixty miles per hour, imagine that there is a simple bathroom scale inserted between your chest and the shoulder strap of your seatbelt. When you hit the brakes and are restrained by the shoulder strap, the strap will push on the bathroom scale and register a force. If it's a one-g deceleration, and you weigh 150 pounds, the bathroom scale should register 150 pounds. The bathroom scale, in this case, is acting as a rather crude accelerometer.

The human body has evolved an extremely sophisticated set of ten linear and rotational accelerometers. And yet some people aren't even aware that they have them, perhaps because, unlike the eyes, ears, nose, tongue, and skin, you can't see them. All those accelerometers are located in the inner ear, right next to the cochlea. You may not be aware of their existence, but it's a cinch you'd know it if they stopped doing their job.

Seismographs, instruments that measure the motion of the ground, are accelerometers. Their best-known application is detecting and measuring earthquakes. Perhaps the earliest seismograph was demonstrated by the Chinese inventor Zhang Heng in A.D. 132. The device consisted of a large bronze vessel with a pendulum inside. When an earthquake shook the ground, the pendulum would swing, tripping one of several levers attached to the ornate dragon's-mouth openings that encircled the vessel. The lever would open one of the dragon's mouths, releasing a ball that would drop to the floor, thus sounding the alarm that an earthquake had occurred. The direction of the earthquake was revealed by which particular dragon's mouth had been opened. The pendulum would swing in different directions depending on which way the ground was shaking, allowing the emperor to send help in the proper direction.

Modern accelerometers are used to control everything from airplanes and missiles to video games. They don't look much like Zhang Heng's seismograph, but conceptually some of them are not very different. Microelectromechanical sensor (MEMS) accelerometers are, as their name suggests, extremely small. Sensor packages smaller than a dime are common. They contain an element, analogous to the pendulum in the Chinese seismograph, that moves when accelerated. The element that moves might be a tiny cantilevered beam with a little mass at its free end. Imagine a minute diving board, shorter and skinnier than a whisker on a day-old beard. When whatever is holding the little diving board gets shaken, the end of the board will deflect, just as a real diving board does when you jump on it. It is that tiny deflection that must be measured, with supreme accuracy, in order to make such an instrument function.

There are several ways to measure such small deflections. In one scheme, the end of the diving board acts as one plate in a capacitor. A capacitor is an electronic device that stores electrical charge. The amount of charge stored in a capacitor is a function of how far apart two oppositely charged plates are held. In a MEMS accelerometer that works this way, as the little diving board deflects, the distance between it and the other plate in the accelerometer changes, and thus the capacitance changes. That tiny change can be accurately measured. The little diving board vibrates up and down, the capacitance in the circuit constantly changes, and the system puts out a signal that can be used in ways that are limited only by your imagination.

## Wii

Accelerometers are showing up in more and more places these days. Consider the Wii. Nintendo's popular entry in the lucrative video game market is refreshingly innovative. Prior to the Wii, most video game controllers used a combination of joysticks and push buttons to control the action on the screen. In a shooting game, the user moves her player around the screen using the joystick. A push button is used to fire the gun. This type of controller is conceptually similar to a computer mouse. Both use a "point and click" system. Various sophisticated mechanisms have been developed that allow computer mice and joysticks to do their jobs, but they all do essentially the same thing: they allow you to point and click. The pointing function translates the motion of your hand, on a mouse or joystick, into the position of something on the screen. Move your hand to slide the mouse across its pad, and the cursor on your computer screen moves with it. The mouse keeps track of its left/right and forward/backward movement in the two-dimensional space of the pad. The computer's operating system takes the signal from the mouse and converts it into the cursor's position. Joysticks do the same thing. While joysticks have undergone numerous innovations and have become quite sophisticated, they retain the original pointing concept described above.

You can play a video game of tennis with a joystick controller, but the motion of the joystick is nothing like the motion of a real tennis racket in your hand. Your player on the screen looks like she is playing tennis, but what you are doing with your joystick won't remind you very much of a real game of tennis. That's where the Wii comes in. The creators of the Wii controller threw mouse/joystick technology out the window and started over, in a largely successful attempt to make video games, especially sports simulators like tennis or boxing, more realistic. The Wii controller looks a little like a conventional TV remote control unit. Inside is a set of accelerometers that sense the rotation of the controller about three axes. You can rock the controller forward and backward, or left and right, or you can twist it like a screwdriver. The controller's accelerometers sense these motions and transmit them wirelessly to the computer that controls the screen. In addition, the Wii controller measures its location relative to the screen using an infrared detector. Thus, when you swing the Wii controller like a tennis racket, the acceleration and position of the controller are transmitted instantly to the motion of your video tennis player's racket on the screen. You "play" Wii tennis in a way that resembles real tennis far more accurately than any previous commercially available video game controller.

The Wii controller is revolutionary, but it is unrealistic in at least one way, and that is the lack of sensory feedback. When you hit a real tennis ball with a real racket, you receive, courtesy of Newton's Third Law,[3] what is known as "force feedback." The harder you hit the ball, the more you feel the force in your wrist, elbow, and shoulder. However, when you "hit" the tennis ball using the Wii controller, there is no force feedback. Your only feedback is visual—what you see on the screen. Some video game joysticks now employ a certain level of force feedback to the user, and I suspect future generations of the Wii controller will do so as well.

Aircraft simulators, such as those manufactured and sold by companies like FlightSafety, are most assuredly *not* video games, although they *are* really cool. Such simulators can cost millions of dollars and are so sophisticated that it is possible for a pilot to become certified to fly a plane she has never flown solely through extensive simulator training. One of the advantages of these simulators is that pilots can be trained to deal with dangerous situations, such as the loss of an engine or a malfunction in the controls, without actually being in any danger.

The instruments and controls in the cockpits of such simulators must be highly realistic. What the pilot sees, hears, and feels inside the simulator has to match reality with as much fidelity as technology will allow. Among other things, the simulators employ feedback mechanisms such that the forces that users must apply to the various controls in a given flight scenario closely match those that would be necessary during a real flight.

Video games are certain to employ this sort of tactile feedback more and more, as the technology becomes cheaper and better, because it is "more realistic," and that's what video game players want. I've never been much on playing video games, but it occurs to me that perhaps the reason they

[3]For every action there is an equal and opposite reaction.

are so popular is not just that they provide an escape from the real world, but that they also provide, more and more as technology improves, a sort of alternative reality—complete with the sort of stimuli, sensations, and perceptions that animate the real world. As the popularity of Wii shows, the ability to incorporate the proprioceptive and tactile senses into a video game is a big step forward.

# Thermal Stimuli

"Wow, that's hot!" someone exclaims, after taking a bite of a tamale or a burrito. "Hot with spice, or hot with fire?" you ask. The stimuli here are either an elevated temperature or a certain chemical ingredient in a food. While these may seem wildly different, they are in fact related, and there's good reason to refer to spicy Tex-Mex food as "hot."

The hotness of chili peppers and the coolness of menthol, as mentioned earlier, are not sensed by the taste buds the way that sweet, sour, or salty foods are. Hot peppers and cool menthol are actually sensed by the same sensory receptors that are activated by hot and cold temperatures.

Chili peppers notwithstanding, thermal stimuli relate to the physical principles of temperature and heat transfer. When you cook breakfast, the skillet is heated by the gas flame or electric element beneath it, causing the atoms in the skillet to vibrate more rapidly. That thermal energy, the intense vibration of the atoms, gets transferred to your bacon and eggs— and cooks them. The same thermal energy, the level of vibration in the atoms, can be sensed by specialized receptors in your skin. If you touch the skillet with your finger, you quickly perceive "hot," and you reflexively withdraw your finger.

The temperature of a gas, such as air, is related to how fast the gas molecules are moving around and bouncing off each other. Blow up a balloon, and tie a knot in the stem. Now gently heat the balloon with a hair dryer. As they warm up, the air molecules inside the balloon will begin to bounce around inside the balloon with more and more velocity. This increases the pressure inside the balloon, causing it to expand. In my career as an engineer, I have investigated several railroad crashes. On one such trip, a veteran railroad worker described to me what happens when a propane tank car is exposed to a fire resulting from a train wreck. Tank cars are those big railroad cars that look like giant hot dogs on wheels. They contain pressurized products such as propane. Imagine an undamaged propane tank car that has the bad luck to be engulfed in the flames spread from another railroad car. The risk is that the propane inside the tank car will get so hot that it will cause the car to explode. I investigated one accident in which shards of steel from an exploding tank car were propelled more than a half mile.

The crew trying to put out such a fire has to work quickly, before the heat from the fire causes the tank car to explode. Just how much time do they have? As the gas inside the tank car heats up, the pressure relief valve on the car eventually opens, blowing off the excess pressure and, for the moment, averting disaster. Then, the valve closes and the pressure inside begins to rise again. A little later, the valve reopens. As the car gets hotter and hotter, the valve opens more and more frequently. Eventually, the car gets so hot that the valve stays open continuously, trying in vain to keep up with the ever-increasing pressure inside the car. Relief valves like this make a lot of racket when they open—not unlike the whistle on an old-fashioned teapot. With the valve continuously open, and the pressure from the intense heat still increasing, the gas whistles out faster and faster, and the pitch of the sound gets higher and higher.

My colleague told me that he and his veteran buddies knew that when the pitch of the whistle reached a certain tone, it was time to get the heck out of there, because the tank car was about to explode. I'm not sure just which musical note we're talking about here, but it certainly made me feel glad that my job was to investigate *after* the incident, not during it.

So that's temperature—a measure of the energy of the molecules in a substance. But it's not just temperature that we sense. The "hot" and "cold" that stimulate us are a complex combination of the temperature of a substance and how fast that substance transfers heat to or from the sensory receptors in your skin. Have you ever jumped into a swimming pool filled with 70°F (21°C) water? I have, and I can tell you it feels so cold it nearly takes your breath away. Yet when I walk outside on a day when the air temperature is 70°F, I feel quite comfortable. Water and air at 70°F have exactly the same temperature, so it's clear that we're sensing something beyond temperature. That something is heat transfer. The same phenomenon is at work in the winter, when you walk, barefooted, from carpeting that you sense as warm to a tile floor, which feels cold. The tile and the carpet are at the same temperature, but when you're walking on the carpet, you are mostly walking on a cushion of air, trapped by the carpet. The overall effect is not unlike the swimming pool versus the outside air.

## Pain Stimuli

Sensory receptors, the cells in the body that convert external stimuli into electrical signals, can be classified in various ways. One useful classification scheme divides up the body's sensory receptors in terms of the stimuli to which they respond the best. There are thus photoreceptors (vision), chemoreceptors (smell, taste, etc.), thermoreceptors (temperature), and mechanoreceptors (touch, balance, hearing, proprioception, etc.). This classification scheme works well enough, until it comes time to describe everyone's favorite sensory stimulus—pain.

The stimulus that a given receptor responds best to is termed the *adequate stimulus*. Receptors that respond well to touch stimuli, for example, respond poorly, if at all, to temperature changes.

Pain receptors are called nociceptors, from the Latin verb *nocere,* to hurt. And pain can be caused by a variety of stimuli.

We just discussed thermal stimuli, so let's use that as an example. If something is too hot or too cold, sensing it is painful. When you stick your finger in a cup of warm (95°F [35°C]) water, it feels warm. But if you stick your finger in a cup of water at 140°F (60°C), it's a different story. Yikes! That hurts! The threshold for pain caused by high temperature varies with the individual, but is generally around 115°F (46°C). The pain threshold for cold temperatures is about 55°F (13°C).

The sensory receptors that signal pain are thus different from the ones that sense temperature. But pain is really a sensation story, more than a stimulus one. The stimulus in the warm/hot water example above is a thermal stimulus; it's just that "too much" of it is perceived as painful. Too much of various other stimuli also cause us to perceive pain. Too much sound pressure (too loud a noise), too much light intensity, too much pressure on the skin, and so on. These are all stimuli that we've already discussed. It's just that the body has evolved defense mechanisms that let us know that stimuli beyond certain threshold levels are harmful and should be avoided. Because stimuli exceeding these thresholds are so dangerous, the body's way of warning us is particularly unpleasant. We call it pain.

The world of the senses is a strange and wonderful mixture of cold, hard science with the warm, soft, fuzzy, and eccentric world of the individual human being. In our tour of the senses, we have arrived at the interface between stimulus and sensation. Welcome to the world of psychophysics— the science of sensation.

# Chapter 4 • The Science of Sensation

I remember having my ears checked by an audiologist at my elementary school. I must have been in the third or fourth grade. Having seated me at a table in an empty classroom, he put some earphones on me and played a series of tones, first in one ear and then in the other. I was instructed to raise my right hand if I heard the tone and my left hand if I didn't. Earphones in place, I raised my right hand repeatedly as I heard a series of tones varying in pitch and loudness. Then, for a while, I heard nothing— so I did nothing. Finally, the exasperated audiologist yanked off my earphones and asked me why on earth I hadn't been raising my left hand. "I dunno," I replied. "How can you tell when you can't hear something?"

This is the problem with the fascinating science of psychophysics. S. S. Stevens defines psychophysics as "the science of sensation." There is an immense and imposing wall that runs down the middle of the domain of the psychophysicist. On one side of that wall resides the scientist with his sophisticated instruments. With those instruments, the psychophysicist measures the various stimuli that we suppose are the inputs to our various senses. On the other side of that wall lies "sensation." Over there, forever just out of reach, is a guy like me, raising one hand or the other in response to a sound, or in some other way responding to whatever stimulus (a taste, a smell, a flash of light, an electric shock) is being visited upon him by our intrepid scientist. No matter how sophisticated the science of psychophysics becomes, it will always be, almost by definition, impossible to completely breach this dividing wall. For in "the science of sensation," there will always be the utterly human world of the senses. You can never take the person out of this equation because that is the whole point of it—to take into account the personal.

But in the science of sensation, we are up against more than just the human problem. For our senses provide us with such a rich variety of different sensations "as to elude complete description," as Stevens notes. "No formula can capture all the richness of the daily sights and sounds and tastes and smells and feelings to which our sense organs admit us."

In order to gain a foothold, we first observe that, when it comes to the senses, there is a clear difference between magnitude and kind. "Sweet is different from sour" (kind), Stevens points out, "although both may vary from strong to weak" (magnitude). The psychophysicist attempts to avoid jumping about from one kind of sensation to another but instead concentrates on areas where there is consensus that a continuum exists, as for example in the sound pressure of a tone with a frequency of 3,000 Hz. We perceive two things, loudness and pitch. Sound pressure (what we measure with instruments) corresponds to loudness (what we perceive). This sensation can be described in terms of "how much." Turn up the volume on your iPod, and it will get louder. Your ears will sense and your brain will perceive more loudness. Frequency (the measured stimulus) and pitch (the perceived effect) are different. The frequency of a sound is clearly measurable and is amenable to psychophysical investigation. We can measure the minimum and maximum frequencies that a subject perceives, and thus we can compare men to women, the elderly to the young, or dogs to cats, in terms of their perceptions. But, while a loud sound has "more loudness" than a soft one, a high-pitched sound is simply *different* from a low-pitched one. It is not a question of "more" or "less."

More research has focused on magnitude scales, such as loudness, than on qualitative ones, such as pitch. One of the most common goals of psychophysics is to define, for a given sense, the relation between *stimulus* and *sensation*. For example, when the stimulus is the sound pressure of a 3,000 Hz tone as measured by a cold, hard instrument, the sensation, what a human being senses through the ears and perceives in the brain, is termed loudness. Stimuli are measured on one side of the psychophysical wall, using instruments. Sensations, on the other side of that wall, are more difficult to measure, since here we must depend on the response of the individual. Anchored firmly in that wall and mediating between the sides are our sense organs.

Psychophysicists, being a tenacious lot, have come up with some clever and sophisticated ways to coax reasonable measurements of sensations out of their living, breathing subjects. They generally do a better job than my erstwhile audiologist. Their attempts to reduce their results to mathematical formulas are not without controversy; but in scientific research, a little controversy is good for the soul. Nevertheless, many psychophysicists believe that stimulus-sensation interactions often follow what is called a power-law relationship, so that the magnitude of a sensation grows as the magnitude of the stimulus is mathematically raised to a power or exponent.

To understand a power-law relationship, let's take a look at the simplest case, in which the stimulus is raised to an exponent equal to one. One such example is a visual sensation known as "apparent length."

In an apparent-length experiment, a person is shown, by means of an image projected on a screen, a straight black line on a white background. The person is then asked to judge the relative length of subsequent lines projected on the screen, compared to the length of the first line. To do this he must "measure them with his eyes" or "eyeball" them. When the results are analyzed, after many subjects have looked at many lines, it is found, on average, that a six-inch-long line appears twice as long as a three-inch line and half as long as a twelve-inch line. The sensation, the apparent length of the line, is proportional to the stimulus, the actual length of the line. This finding appears reasonable, but in fact a result like this is quite unusual.

More often, researchers discover that the magnitude of the stimulus increases much faster than the magnitude of the sensation. It has often been observed that this is a very good thing. Because sensation increases far more slowly than stimulus, we are able to safely, without damage to our sense organs, experience sensations over an enormous range of stimuli. Sound is a good example. The roar of a jet engine, heard at a distance of, say, 100 feet, might be a million million ($10^{12}$) times more powerful, as measured in watts per square meter, than the buzz of a mosquito at 3 feet. But our sensation of loudness is not a million million times as great. It is only, roughly speaking, about twelve times as great. A similar relationship is observed for the eyes when sensing the brightness of a light. Our senses are thus well suited to our natural environment, with its wide range of stimuli. This is hardly a surprise. After all, that's how evolution works. Because most of our sensations operate this way, a flash of lightning does not blind us, nor does the thunder that follows deafen us. A few seconds later, we can still both see and hear the mosquito about to land on our forearm.

How, you might ask, was it determined that the jet engine in the above example is perceived as twelve times louder than the mosquito? Well, that's psychophysics. Let's look at loudness in more detail, since it is one of the best known and most studied stimulus-sensation relationships. Loudness, to be sure, is the sensation in this case, on the far side of the psychophysical wall. Acoustic stimuli can be measured in terms of energy flow, in watts per square meter, or in terms of sound pressure. The metric unit for sound pressure is the Pascal (Pa). One Pa is equal to one Newton per square meter. The equivalent U.S. unit is pounds per square inch.

Sound pressure level is often presented on the familiar, but confusing, decibel scale. The decibel is misunderstood in so many different ways that it's difficult to count them all, much less describe them. First of all, we don't *hear* decibels, we *measure* them. They are measures of sound pressure (stimuli). It turns out that sound pressure level, as represented in decibels, does a good job of approximating what we hear—the sensation of loudness. Another problem with the decibel is that it utilizes a logarithmic scale, a mathematical concept many of us suspect was designed to strike terror into the souls of high school math students. Since the decibel scale is logarithmic, a sound pressure of 70 decibels is ten times as great as a sound pressure of 60 decibels, which is in turn ten times as great as 50 decibels, and so on.

Another common misconception regarding the decibel is that it is exclusively an acoustical measurement. Pretty much anything can be measured in decibels. It is a unitless scale, involving the ratios of numbers. You could measure people's heights in decibels if you wanted to. That wouldn't be particularly useful, since people's heights don't vary all that much—at least not compared to the sound pressures that our ears are capable of hearing. Decibel scales are usually devoted to measuring things that vary over many "orders of magnitude," where an order of magnitude is one multiple of ten.

The decibel scale was invented in the 1920s to quantify the energy loss per mile when transmitting a signal along a telephone line. The *bel* in *decibel* is in honor of Alexander Graham Bell. To measure transmission loss, sound pressure, or anything else in decibels, you start with a "reference value." For sound pressures, the reference value is often taken as $20 \times 10^{-6}$ Pa in pressure units. This is generally considered the threshold of hearing, for a sound with a frequency of 3,000 Hz. For sound pressures below that, the average person hears nothing. Above the reference value, the sound gets increasingly louder.

By mathematical definition, the reference value always lies at zero decibels. To turn any measurement into decibels, do the following: First, divide your measurement by the reference value. Take the base ten logarithm of that ratio, then multiply the result by ten. What you get are decibels. From this, you can see why the reference value always lies at zero decibels: dividing the reference value by itself gives a result of one, and the logarithm of one is zero, and it's still zero after you multiply by ten. Now, turn up the volume so that the sound pressure level is ten times as great as the reference value. This gives a value of 10 decibels, since the logarithm of ten is one, which you then multiply by ten. Crank it up once more, until the sound pressure level is one hundred times the reference value. That gives you 20 decibels, since the logarithm of one hundred is two. And so on. Every time you make the sound pressure level ten times as great, you add 10 decibels.

One reason the decibel is used in acoustics is that it is such a good match in terms of stimulus and sensation. Relating stimulus to sensation is the goal of psychophysics, after all. But this points out yet another problem with the decibel in acoustics. Because it is a logarithmic scale, and thus roughly mimics loudness, small changes in the number of decibels measured represent relatively large changes in the loudness that is perceived.

Psychophysicists are often interested in the smallest change in a stimulus that a person can perceive, sometimes referred to as the "just noticeable difference," or JND. Getting a prescription for eyeglasses is a good example. The eye care professional asks you a lot of questions about what you can see and what you can't while your eyes are looking through a series of different lenses. The JND is an important part of these sessions, although I've never heard it called that in this context. As the eye care professional zeroes in on the right prescription for me, he eventually arrives at two sets of lenses that look just the same to me. I can't tell which pair makes my vision sharper. At that point we're within the JND and so must be pretty close to the right prescription.

Here's another JND example. While channel surfing, you chance upon a program that catches your eye. But the volume isn't loud enough. How many times must you click on the "volume up" button on the remote control before it sounds noticeably louder? One click? Two clicks? Three? Whatever the result, that is your JND.

The just noticeable difference in sound pressure level is often reported as 3 decibels, although it varies from one person to the next and is also a function of both sound pressure level and frequency. And how much louder does a sound have to be before the average listener perceives it as "twice as loud?" That depends on lots of things, such as the frequency of the sound. Beyond that, it depends on whose results you believe. Somewhere between a 6- and 9-decibel increase in sound pressure level is generally perceived as "twice as loud." So the just noticeable difference is around 3 decibels, and the twice-as-loud increase could be as small as 6 decibels. This gives you some idea of just how close together things are on a decibel scale.

Several years ago I helped a group of students on an acoustical engineering project. Their goal was to modify an existing leaf blower to make it quieter without reducing its performance. For me, the leaf blower is the bête noire of our modern noise-polluting society, so I leaped at the chance to help out. Their finished product, to my ears, was a great success. Using sophisticated sound-measurement instruments, the students first identified and characterized four different noise sources on their small gasoline-powered machine. These were the discharge tube (where the high-velocity air comes out to blow the leaves), the engine exhaust, the engine inlet, and the inlet to the fan that pushes the air down the discharge tube. They then designed four different solutions, one for each of those four noise sources, and adapted them to the machine. When they were through, the students were able to show that their modifications reduced, by about 6 dBA, the noise produced by the leaf blower running full-throttle at a distance of fifty feet, with no change in leaf-blowing performance. The A in dBA refers to the "A level," a common measure that is essentially a weighted average of all the noise produced by the leaf blower across all the different sound frequencies. The dBA measure is thus a means of getting around the frequency-related issues inherent in sound measurement and boiling the results down to a single number.

When, however, the students gave presentations on their project, which won several awards, I noticed that this 6 dBA improvement generally was perceived as not all that impressive by the various audiences, some technically trained, some not. Six just isn't a very big number. Even though the students always hastened to add that "a 6 dBA reduction is often perceived as half as loud," the audiences seemed unimpressed. Things changed, however, when they ran the machines for an audience with and without the noise-reducing modifications. Six dBA is just a number when it's written on a page, but the ear doesn't lie when it gets the chance to really hear what's going on.

# Part 2 • Sensation

As a result of a childhood disease, Michael Chorost's hearing was severely damaged. Powerful hearing aids in both ears enabled him to make good use of what remained of his hearing, and that, along with the ability to lip-read, allowed him to function in the world of the hearing for several decades. One day when he was traveling on business, his remaining hearing ability suddenly disappeared, a fact he realized only after trying several new sets of batteries in each of his hearing aids. Not long thereafter, Chorost became one of the early recipients of the cochlear implant, a miraculous device that can restore a sense of hearing to many profoundly deaf individuals.

The cochlear implant is the first artificial device that actually replaces a set of sensory receptors in the human body. These days, what Michael Chorost hears is created by a computer, which takes in sound from a microphone and generates electronic signals that are fed directly to Chorost's brain. The cochlea, the magnificent organ of the inner ear that normally transforms acoustic pulses into electronic signals, is completely bypassed.

When we look at each other, most of the sense-related equipment we see (the pupil and the iris of the eye, the external ear, the nose) is part of our sensory system's "signal conditioning hardware." This hardware receives stimuli and conditions or prepares it to be analyzed by the body's various sensory receptors. Most of the eye exists to condition light by both aiming it at and focusing it on the one part that really matters: the retina, the interface between eye and brain. The sensory receptors, known as rods and cones, are located in the retina. In similar fashion, the external ear channels sound waves into the middle ear, where they are amplified before entering the liquid-filled cochlea, where the sensory receptors, known as hair cells, are located. Our sensory receptors are where sensation really takes place. They are the psychophysical boundary. On one side, all is stimulus; on the other, electronic signals.

Not many years ago, the sensory system in a brand new car was very limited. The driver had to rely on her own sensory system for all aspects of vehicle control and for most aspects of detecting the proper operation of the car's various subsystems, such as steering and braking. Today's cars, in contrast, might be thought of as a large collection of high-tech sensors rolling down the road in close formation. Some of the sensors help ensure safety during driving, while others help prevent serious maintenance problems. There are sensors that monitor wheel speed and help keep your wheels from locking during braking. There are temperature sensors that monitor engine oil, engine coolant, and exhaust-gas temperature; others measure the temperature inside and outside the vehicle. Pressure sensors monitor engine oil. All these modern automotive sensors have this in common: they measure an external stimulus and transform it into an electronic signal, which is then passed on to a computer in a process analogous to the human sensory system.

In the good old days, cars had only one speed sensor. It mechanically measured the speed at which the transmission was rotating and from that was able to display road speed on an analog speedometer. These days, most cars include a variety of speed sensors, including one on each wheel, in order to facilitate the car's antilock braking system (ABS). These latter sensors are so responsive to changes in wheel speed that they can, on some vehicles, alert the driver to a tire with low air pressure. Low pressure reduces a tire's diameter slightly, which makes that wheel rotate a little faster than the others.

The exhaust from a car's engine is a complex mixture of dozens of very hot chemicals. A sensor in the exhaust stream measures oxygen content, which the engine's computer uses to control the air-fuel mixture being fed to the engine. This optimizes engine performance and minimizes harmful exhaust gases.

When a car's airbag explodes, the gases generated by the blast fill the bag in a fraction of a second, preventing your head from slamming into the dashboard or steering wheel. The explosion is set off by accelerometers that sense a rapid change in the car's velocity, as during a crash.

There are other sensors as well, including, on some cars, photocells that turn the headlights on and off automatically, moisture sensors that activate windshield wipers, and audio sensors that respond to voice commands.

All these human-made sensors have started to make the modern car evoke the human body, which features its own collection of highly sensitive, hyperspecialized sensors. As sophisticated as the system of sensory receptors in modern cars have become, they have many miles to go to match the system with which most of us were born.

Sensory receptors are specialized cells that produce electrical signals in response to the type of stimuli they are optimized for. Sensory receptors can be categorized various ways. One system classifies them according to the stimuli to which they are most sensitive. We thus have chemoreceptors, such as those for smell and taste. Photoreceptors in the retina respond to light. Thermoreceptors respond to temperature. Mechanoreceptors form the broadest subgroup, responding to various types of force and motion. Some mechanoreceptors respond to touch, and others monitor muscle length and tension. And there are the amazing hair cells in the cochlea for hearing and in the balance system for tracking the acceleration and the position of the head. The neatness of this scheme is upset a bit by pain receptors, since these can respond to a little bit of everything, including various chemical, thermal, and mechanical stimuli. Pain receptors are thus often placed in a category by themselves.

# Chapter 5 • Vision

If a poll were taken, vision would surely be the hands-down choice as our most vital sense. Perhaps the brain agrees, since it allocates more volume for the visual cortex than for the rest of the senses combined, although other animals aren't always wired this way.

Vision is also the only sense we can turn off. When we close our eyes, we come pretty close to shutting out all the electromagnetic waves that stimulate the retina. A small dose of radiation does get through. When we are sunbathing, our closed eyes can tell us when the sun goes behind a cloud. But that's about it. Vision is the only sense with a built-in mechanism for blocking stimuli from its sensory receptors. You can hold your nose or cover your ears, but these are relatively ineffective and lack the automatic nature of the eyelids. Try holding your nose and then going to sleep. It doesn't work. Smell, hearing, balance, temperature sensing, and pain are on duty around the clock, alerting us to danger if nothing else. So why do our eyes get the night off when we sleep? It's possible that our automated eyelid system evolved because so much of our brain's computing power is devoted to vision. When we sleep, lots of the things our brains are occupied with during waking hours (the job, the kids, the house) get turned off, and so does vision. But we're getting a little ahead of the story.

## My First Glasses

I got my first pair of eyeglasses when I was six years old and a first grader in Providence, Rhode Island, in 1963. My teacher had noticed me squinting at the blackboard and had called my parents. Looking out the car window, wearing my new glasses for the first time, I was delighted to discover there were all sorts of things out there with words on them, such as road signs and billboards. Having just recently begun a lifelong love affair with the printed word, I annoyed my parents and pestered my younger brother for weeks by reading aloud everything I could now see outside the car window.

The intervening years, nearly fifty of them, have given me plenty of time to get used to seeing everything pretty clearly with the aid of my glasses. But things are changing. In my midforties, I first began to notice a slow, frustrating, but utterly normal process: I began losing the ability to focus on things close up. This process, leading to the condition known as presbyopia, begins at a much younger age, but most of us don't notice it until we're in our forties. "Put up with it as long as you can," my ophthalmologist said, "and then we'll get you some bifocals." Putting up with it is more annoying than I expected, but I have thus far resisted the temptation to avail myself of the bifocals, that classic symbol of late middle age.

## IR Vision in Snakes

Before we discuss what we *can* see, and how we see it, let's consider some of the things we *can't* see—things like x-rays, ultraviolet radiation, infrared radiation, microwaves, and radio waves.

Our ears can't hear all the frequencies present in sound waves, and our eyes certainly can't sense all the frequencies across the electromagnetic spectrum. We can't even see all of the frequencies present in sunlight.

Electromagnetic waves longer than 700 nanometers or shorter than 400 nm are invisible to humans. Certain other creatures can sense radiation outside the visible range, however. Some snakes can sense infrared radiation, known as heat waves, which have wavelengths up to about 400 times longer than those of visible radiation. "Sense" is a better word than "see" for this reptilian ability, since snakes use a different sense organ, called a pit organ or a pit hole, for this task, and not their eyes. Technology allows humans to use their eyes to sense radiation outside the visible range, but only with the aid of instruments such as, for example, "night vision" goggles, with which the eyes can detect infrared radiation.

Two groups of snakes, pit vipers and boids (the latter group includes boa constrictors), are equipped with a sensory instrument that appears to be found nowhere else in nature. The pit hole is a shallow depression located between the nostril and the eye in these snakes. It allows the snakes to detect infrared radiation, an ability they put to good use hunting prey in the dark.

All objects emit radiation based on their temperature. If an object is hot enough, the radiation it gives off is visible to humans. Think of the filament of an incandescent electric light bulb or a red-hot piece of steel. *Incandescent* comes from the Latin *candescere,* to glow or become white. Cooler objects give off radiation too, but at longer wavelengths than those visible to humans. The infrared radiation emitted by such objects can be measured by an infrared thermometer, as shown in figure 5, or by the pit hole sensor of a boa constrictor.

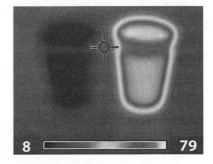

*Fig. 5.* Conventional (*left*) and IR photos of two cups of water. In each photo, the cup at left contains cold water (46°F) and the cup on the right contains water at 174°F. (Photographs by the author)

The pit hole is a tiny hemispherical depression roughly 1 millimeter deep and 1 millimeter in diameter. The cells at the bottom of the pit are extremely sensitive to infrared radiation. They can detect temperature differences on the order of thousandths of a degree Celsius.

The pit hole operates on the same optical principle as a pinhole camera. Pinhole cameras have been around for a long time. The Arab physicist Ibn al-Haytham wrote about them in his *Book of Optics* in A.D. 1021. These simple cameras retain a small but devoted following among artists and others even today. The one I made when I was a kid didn't work very well, as I recall. A pinhole camera utilizes a tightly closed box or other container. A tiny hole, not much bigger than a pencil point, is the only way light can enter the container. Light passes through and, because the hole is small, is focused on the opposite side of the container, where it creates an inverted image on a piece of film. The smaller the hole, the sharper the image. But small holes admit very little light, so long exposures are required to achieve reasonable images. Pinhole cameras are thus generally suitable only for still-life photography.

The pinhole-camera problem is very much the problem with the pit holes in snakes. The snake utilizes the pit hole to hunt its prey in the dark. But the creature the snake is after tends to be moving, not still. To be of value, the pit hole has to be able to detect, for example, a rat scurrying across the landscape a few feet away. The pit hole sense organ is thus presented with a conflicting set of objectives. On the one hand, it must admit enough infrared radiation so the snake can sense the presence of the moving rat. This requires the opening of the pit hole to be relatively wide. On the other hand, the IR image needs to be focused well enough so that the snake can successfully strike its target. This requires the opening of the pit hole to be relatively small.

An optimally designed pinhole camera employs a hole diameter about 1 percent of the distance from the hole to the film. Thus, a camera made from a cylindrical Quaker oatmeal container, which has a diameter of about 5 inches (125 millimeters), would need a hole in the cylindrical wall of the container with a diameter of 0.05 inch, or 1.25 millimeters. The film would then be placed on the opposite interior wall, 5 inches away. A boa constrictor's pit hole, in sharp (or rather, fuzzy) contrast, has a hole diameter of about 1 millimeter and a hole-to-"film" distance of 1 millimeter. This one-to-one ratio guarantees that the IR image created by the pit hole will be fuzzy in the extreme.

Although it is far more sophisticated, the human eye behaves like a pinhole camera in at least one respect. In bright light, the iris at the front of your eye closes so as to admit light through the pupil via a tiny hole. You can verify this. Look at your eye in a mirror, then shine a flashlight on that eye. Your will see that your pupil constricts, admitting less light into your eye. This is the pinhole effect, and it sharpens the focus beyond what your eye's flexible lens can do all by itself. The older I get, the more I appreciate the advantages of strong lighting. The lenses in my eyes are no longer as flexible as they were, and it gets harder and harder to focus on things close up. Strong light helps because of the pinhole effect.

For the snake, though, the need for a wide opening in the pit, to let in lots of radiation, must be more important than the creation of a sharp image. If the pit hole were optimized like a pinhole camera to create a very sharp IR image, it would be of little use to the snake. That rat is not going to hang around long enough to let enough IR enter the pit hole to allow its image to be faithfully recorded by the snake's sensory apparatus. The pit hole is a quick and dirty way for the snake to get a fuzzy image that something warm is wandering by in the dark. The snake's goal is not to admire nature in all its glory, but rather to get something to eat.

It is also possible that the raw IR image created by the crude optics of the pit hole is enhanced through a process of "image reconstruction" in the snake's brain. A team of German physicists led by A. B. Sichert have investigated this idea and have created a plausible model for how it might work. Experiments have shown that snakes with pit holes can locate heat sources to within an angle of 5 degrees. The field of view of the pit hole spans an angle of about 100 degrees. Given that the pit hole membrane consists of a paltry 40 by 40 grid of IR-detecting cells, an accuracy within 5 degrees is a remarkable performance.[1]

Image reconstruction would allow the snake to do a better, more accurate job of locating prey in the dark than would seem possible given (a) the very fuzzy optics of its "pinhole camera" and (b) the relatively small number of IR-sensitive cells it possesses. We've all seen movies and TV shows where a fuzzy photograph is "image enhanced" on a computer to produce a clear portrait of the bad guy committing a crime. Hollywood's version of image reconstruction may be exaggerated, but the basic idea is correct: to make up for missing or inaccurate information in an image through sophisticated software, resulting in an enhanced image. Image reconstruction is important in various technologies besides law enforcement, including medical imaging techniques like MRI.

In addition to an IR-sensing pit organ, a snake has eyes and can sense visible light as well as infrared. Unless the darkness is truly profound, the snake can use its normal vision to enhance the information obtained using its IR pit hole. Combining a fuzzy IR image with a faint visual one appears to be enough to allow the snake to accurately locate and strike its prey.

## Human Vision

The eye is tasked with transforming visible light waves into electrical signals, ready to be processed by the brain. In short, this is how the eye does it. When light strikes the eye, the eye regulates how much light can enter and then focuses it on the inner, back surface of the eyeball. Muscles control the focusing mechanism, while other muscles aim the eyes in unison. Inside the eye, the light is absorbed by four different kinds of receptors that are optimized for wavelength and intensity. The light receptors transform the energy in the light into signals that are carried into the brain.

---

[1]In contrast, the human eye contains tens of millions of light-sensing rods and cones.

There "visual perception" occurs. The details of that process we leave for part 3, "Perception." For now, let's focus (no pun intended) on the details of the eye itself. The marvels therein can take your breath away.

The eye is often compared to a camera, and in many ways the comparison is apt. Like a camera, the eye includes a mechanism for regulating the amount of light that enters through the front of the eye. That opening is called the pupil, and the amount of light that enters is controlled by the iris. A cross section of the human eye is shown in figure 6. Like a camera, the eye contains a lens for focusing the light. But the lens in the human eye works differently from the lenses in most cameras. In the human eye, the lens changes its focus by changing its shape, whereas in a camera a fixed-shape lens moves closer to or farther away from the film or, in a digital camera, the photosensor. There are some animals, some fish, for example, whose eyes focus by telescoping a fixed-shape lens in camera-like fashion.

Most of the structure of the ear exists for one reason: to stimulate the motion of the hair cells in the cochlea in the inner ear. In the same way, most of the structure of the eye exists to focus light on the retina. Up to but not including the retina, the eye is essentially a signal-conditioning

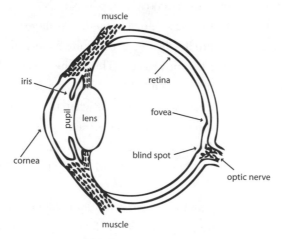

*Fig. 6.* Cross section of the human eye.

device. The human eye is roughly spherical and about 25 millimeters (about an inch) in diameter in the average adult. The mass of the eye is only about 8 grams—about the same as a half tablespoon of water. Think of the eye as a three-layered hollow rubber ball filled with clear jelly. The outer layer of the ball, called the fibrous layer, consists of dense connective tissue that both protects the delicate apparatus within and maintains the eye's spherical shape. Inside the fibrous layer is the vascular layer, whose blood vessels supply the eye with oxygen and carry away waste. The nervous layer, the innermost of the three layers, includes the retina.

## The Hardware of Vision

### The Cornea and the Lens

The cornea, at the front of the eye, and the lens, just behind, act as a compound lens to focus light on the back of the eye, the retina. The retina might be thought of as the film of the eye, although the comparison to a modern photosensor is perhaps more apt. In a camera (the old-fashioned kind, using film) light is focused through the lens on the film, where a chemical reaction takes place that records the image. In the eye, the image must be focused on the retina, for it is there that the miraculous process of transforming light into electrical signals begins.

Imagine that your eye is the size of a basketball—about nine inches in diameter. Unlike a basketball, the eye is not quite spherical. At one end there is an almost blisterlike protrusion. This is the cornea. The cornea has two main roles: to let in all the light that we see and to help bend (or refract) it so that the light rays will be in focus when they strike the retina. In this second role, bending the light, the cornea works in concert with the flexible lens that is located behind it. If your eye were the size of a basketball, the cornea would extend outward by a half inch or so beyond the arc of the sphere.

When I look at my eyes in the mirror, it's not evident that the cornea protrudes from the spherical rest of my eye. I can't tell, because I'm looking at my eyes straight on. But if I hold a second mirror at nearly a right

angle to the first, and then get really close to the first mirror, my nose nearly pressed to the glass, I can observe, in the side mirror, that my eyeball does indeed bulge outward. But it's very difficult to see the cornea itself, in either mirror. What we mainly see, looking at an eye, is the colorful iris behind the cornea. For one thing, the cornea is the essence of transparency, all the better to allow light to enter the eye. For another, it's quite thin, varying from about 0.5 millimeter (0.02 inch, or roughly the thickness of five pages of this book) at the center of the cornea to close to 0.8 millimeter near the edges. This thickness variation is what accounts for the cornea's prodigious light-bending ability.

When light travels from one medium to another, say, from air to water, the light waves are bent, because light travels at different speeds in the two media. This is the familiar refraction phenomenon. The trout you see swimming in a clear stream is not exactly where your eyes tell you it is. Water bends the light waves and can fool your eyes. A magnifying glass bends or refracts light, too.

And so it is with the cornea. Imagine someone were shining a flashlight through the cornea of a basketball-sized eye. The cornea would help bend the light rays so that they would be in focus, at a point on the inside back surface of the basketball opposite the cornea. The cornea is something like a contact lens, permanently bonded to the outside of your eye. The light-bending power of a contact lens is fixed, and that of the cornea is nearly so. The ciliary muscles that change the shape of the lens also bend the cornea to some extent, changing its radius of curvature and thus its light-bending ability. LASIK and other types of refractive eye surgery involve permanently changing the shape of the cornea by removing part of it, such that the cornea, after surgery, bends light differently than it did before surgery.

The maximum light-bending ability of the human eye is about 60 diopters (the diopter is defined a little later). The cornea accounts for around 40 diopters, or two-thirds of that total, while the adjustable lens behind the cornea takes care of the rest. Since the cornea's refractive ability is nearly fixed, LASIK notwithstanding, it is largely the role of the lens to adjust its own shape and thus allow us to focus on objects at varying distances. The lens, and to a lesser extent the cornea, change shape through the contractions and relaxations of the ciliary muscles.

When the ciliary muscles relax, the lens assumes a relatively flat shape and focuses on objects that are farther away. When the muscles contract, the lens becomes more round, and the cornea bulges out a little bit more, allowing the eye to focus on things close up.

There are various theories of how the cornea achieves its transparency. But it isn't always transparent; diseased or otherwise damaged corneas are a significant cause of blindness. The cornea is the window at the front of the eye. If the transparency of that window is compromised, everything behind it must suffer.

The first cornea transplant was performed in 1905. It was one of the first transplants of any kind. Today, healthy corneas are harvested from accident victims and others who have agreed to be organ donors. Eye banks collect and distribute donated corneas. The surgery is done by ophthalmologists, often on an outpatient basis. Artificial corneas do exist, and they work well, but they are typically utilized only after a conventional donor transplant has failed or when a donor transplant is deemed inadvisable. One reason artificial corneas are not prescribed more often is that they are extremely costly.

The eye, especially if it is young, changes focus very quickly, but not infinitely so. On a nice sunny day, try glancing from the book you are reading out the window to the horizon and then back to your book. If you concentrate on this task, you will probably notice that each successive image takes just a fraction of a second to focus. This is the lens trying to keep up with the work of focusing the eye, first on a close-up object, then on one far away.

With age, alas, the lens loses some of its elasticity, a trait it shares with other parts of the body. When the ciliary muscles contract, the lens can't get as round as it used to. The unaided eye can thus no longer focus as it once did on close-up objects, such as books and computer screens. This condition is called presbyopia, and it is why most people, beginning in late middle age, must wear bifocals or at least reading glasses. Presbyopia means "elder eyes." It comes from the Greek *presbys*, for elder, and the Latin *opia*, related to the eyes. But the ability to focus on close-up items begins deteriorating at a surprisingly young age. Children under ten can typically focus on items as close as two inches from their eyes. By the time a person reaches age twenty-five, that distance has doubled to four inches. We generally don't notice the change until around age forty-five, when items closer than eighteen inches begin to go out of focus. This deterioration starts to level off at around age sixty, when most of us have difficulty focusing our eyes on anything closer than about three feet. Figure 7 quantifies the change in refractive ability that the aging eye undergoes.

Night vision also degrades with age. Our ability to see in darkened conditions deteriorates, on average, twice as fast as does our ability to see in normal light. In addition, the pupil's ability to dilate and let more light enter the eye decreases with age. On average, a twenty-year-old eye admits, under conditions of darkness, about sixteen times as much light as an eighty-year-old eye. My seventy-seven-year-old mother sees quite well for a woman her age. She plays an excellent game of tennis, without glasses, and gets around town just fine driving her car. At night, though, she is

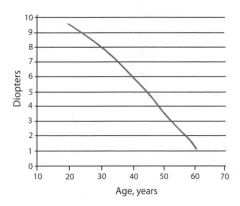

Fig. 7. Average deterioration with age in the maximum light-bending ability, in diopters, of the lens of the human eye. (Adapted from Atchison and Smith, *Optics of the Human Eye*)

extremely reluctant to get behind the wheel. I'm in my fifties, and I'm learning why Mom is so hesitant to drive at night, since I have more difficulty seeing at night than I used to.

The eye's lens yellows with age, changing and degrading our color vision and our ability to see in dim light. Paco Underhill notes that this influences how the elderly shop and what they buy, with implications for advertisers and retailers everywhere. Other age-related changes to the lens and the cornea result in increased sensitivity to glare and cause other degradations in the quality of vision. If all this sounds a bit hopeless, take heart, because there are some relatively simple things anyone can do to mitigate the effects of aging eyes. If you wear glasses, antireflective lens coatings are a must. These reduce the amount of light that bounces off your eyeglass lenses from around 10 percent to less than 1 percent. The less light bounces off your glasses, the more light enters your eye, which is important for the aging eye.

Many of the things that one can easily do for the aging eye relate to lighting. The type of lighting (fluorescent versus incandescent, halogen, or LED), the location and orientation of lighting, shielding and lens coverings, and other factors are all important. The Lighting Research Center at Rensselaer Polytechnic Institute provides detailed lighting recommendations for the elderly on its Web site at www.lrc.rpi.edu.

## The Retina

The eye is not hollow. In addition to the iris, the colored part of the eye that controls the size of the opening in the pupil, the lens, and a few other items, the inside of the eye is filled with a transparent jellylike substance known as the vitreous humor. Light passes through the lens, which is suspended by a network of muscles in the jelly, and it is then focused on the back of the eyeball, on the inside surface. That surface is the retina. The organ of Corti in the ear, as we shall see, is sometimes referred to as the seat of hearing. In analogous fashion, the retina might be called the seat of vision. The retina is precisely where electromagnetic energy, in the form of visible-light waves, is transformed into electrical signals and sent on to the brain.

The retina covers about three-quarters of the inside surface of the eye, but it is not completely smooth or uniform. There are two particularly unusual regions on the retina. One of these is the optic disk, a small spot where the optic nerve attaches to the retina. This spot is devoid of photoreceptors and is literally blind. Having two healthy eyes allows us to compensate for the blind spot that each possesses. Each eye has its own blind spot, but since they don't overlap, we have a continuous field of vision with both eyes open. Even with one eye closed, we don't perceive a blind spot. Various parlor tricks can reveal the presence of this blind spot. We'll save ours for part 3, "Perception." The brain simply fills in the blind spot in our visual field with a reasonable guess as to what is likely to be there. This "perceptual completion" has profound implications for how vision and the other senses work.

The other atypical region on the retina is found at its center, directly opposite the lens. At that point there is a tiny dip in the retina, called the fovea, which is Latin for *pit*. In the fovea, tens of thousands of photoreceptor cone cells are packed in with maximum density. There are very few rods in the fovea, and at its very center there are no rods at all.

The fovea is almost entirely responsible for the sharp color vision most of us are lucky enough to possess. On a basketball-sized eye, the fovea would be about 9 millimeters in diameter, about half the diameter of a dime, or not quite big enough to type the word *fovea* on it on the inside surface of the basketball. The actual fovea is much smaller—roughly 0.5 millimeter across, or smaller than the hole in the middle of this *o*. That small spot can be called the seat of vision within the seat of vision that is the retina.

The fovea is located in the center of the region of the eye known as the macula. The macula is best known through the term *macular degeneration*, for one of the most common causes of blindness, especially among the elderly. The macula of the retina, where the fovea is located, is so crammed with photoreceptors that there is no room for blood vessels—the macula has no blood supply of its own. When the light is really bright and the macula is working very hard, it operates under hypoxic (lack of adequate oxygen) conditions. Of the two common types of macular degeneration, dry and wet, the latter is related to the blood supply that feeds the macula.

Much of the external machinery of the eye, the muscles and the connective tissue, exists to rotate the eyeballs in unison such that the image of whatever the brain wants to look at is centered on the two foveae, one in each eye. This is no mean feat, since the tiny fovea sees, literally, only the central *two degrees* of your field of vision. The visual arc seen by the fovea is roughly equivalent to that of a quarter held at arm's length. Processing the information from that tiny patch of real estate takes up more than 50 percent of the computing power of the visual cortex of the brain. Roughly three-quarters of the inside surface of our eye is covered by photoreceptors capable of converting light into electrical signals. Yet only that single half-millimeter spot, the fovea, is capable of producing the really high-quality visual image that most of us couldn't imagine doing without.

Lots of exercises have been conceived to demonstrate just how restricted the visual field processed by the fovea is and how inferior the images are from elsewhere on the retina. One such example will suffice. Lay the book flat on the table, and put a quarter face up right in the middle of a page. Focus on the quarter, so that you can comfortably read the lettering above Washington's head. Now, *while maintaining your focus on the quarter,* try to read the words to the left and right of the quarter. I find it impossible to identify more than a word or two in either direction, and that is typical. My eyes tell me those are words over there, but I can't read them while I'm focusing on the quarter. The urge to cheat, however, is almost irresistible, and as soon as I do so, by shifting my focus left or right, the words are instantly identifiable. We do this, we shift our focus so effortlessly that it seems as though everything is in focus all at once. But it isn't true. Only a tiny patch of our visual world is presented to our brain at the highest quality of which we are capable. All the rest is, literally, visibly

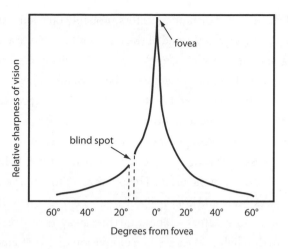

*Fig. 8.* The relative sharpness of vision, in normal light, of the human retina, shown in relation to the position of the fovea, as measured in degrees along the arc of the retina. Note the blind spot about 15 degrees to the left of the fovea, indicating that this is the right eye.
(Adapted from Nolte, *The Human Brain*)

poorer. Figure 8 shows the relative reduction in visual acuity that occurs away from the fovea.

## THE RODS AND CONES

When you wake up in the middle of the night for a bathroom visit, unless your room is profoundly dark, your eyes allow you to navigate relatively well across your bedroom and down the hallway. Once inside the bathroom, you turn on the light, and the first few seconds of illumination are blindingly bright. But your eyes adapt very quickly. When you turn off the light, however, you may wish you had a guide dog for the return trip to your bed. The night vision that was so useful on the way to the bathroom is absent, leaving you helplessly in the dark for the return trip. Coming from darkness, our eyes adapt to light within seconds, but going from bright light to darkness is another story. Night vision increases markedly for the first thirty minutes or so after you shut off the lights, and it isn't fully developed for up to an hour. Since natural light fades gradually at the end of the day, our distant ancestors probably enjoyed a smooth transition into night vision at the end of the day. But then came artificial light. Explaining phenomena such as night versus day vision requires an exploration of our visual sensory receptors—the rods and cones.

The process by which light is transformed into electrical signals is called phototransduction, and it takes place in the retina. The retina contains millions of photoreceptor cells, whose job it is to absorb light and begin the process of passing along information about that light, such as its color and brightness, to the brain. The photoreceptors in the retina are of two basic types: rods and cones. Their names approximate their shapes. Rods are found in greater numbers farther away from the fovea. They help us see in dim light. Cones are for color vision and are found in greatest concentration in the fovea. They are of three types, depending on the color (wavelength) of light they best absorb. There are S (short-wavelength), M (medium-wavelength), and L (long-wavelength) cones. These are sometimes referred to as blue, green, and red cones, although those names are not particularly accurate. The maximum sensitivities of the three types of cones occur at wavelengths of about 440 nm (blue-violet) for S cones, 540 nm (yellow-green) for M cones, and 580 nm (orange-yellow) for L cones; these numbers vary depending on which expert you consult.

These three types of cones, each optimized for light of a different wavelength, make our color vision possible. Recall our discussions of additive and subtractive color theory in part 1. Whether we are mixing pigments, as with paints, or adding different-colored beams of light, we can create millions of colors from one end of the visible spectrum to the other. In like fashion, having three types of cones, each absorbing from a different part of the spectrum, is the physical basis for color vision.

Cones function best at normal levels of brightness. This is why, in the middle of winter, I sometimes show up for work wearing one black sock and one blue one. In the near-darkness of my clothes closet, I can see well enough to know that I'm holding two dark-colored socks in my hand, but there isn't enough light for my cones to help me distinguish the colors.

Rods come in only one type. They are optimized to absorb blue-green light with a wavelength of about 500 nm, but since there is only one type of rod, there is no color-mixing effect as with cones, and rods can't discriminate different colors, hence my wintertime sock dilemma. What rods do have, relative to cones, is incredible *sensitivity* to light. Rods are more than a hundred times more sensitive to light than L or M cones and nearly ten thousand times as sensitive to light as S cones. A single photon, the tiniest individual packet of energy contained in light, is said to be able to stimulate a rod cell. Astronomers figured out a long time ago that they could better detect faint stars and planets through their telescopes if they observed these objects not by looking at them directly, but by looking out of the corners of their eyes, as the saying goes. By avoiding the fovea, they were bringing a much higher percentage of highly light-sensitive rods to bear on their obscure targets, and relatively few low-sensitivity cones.

Because S cones are less sensitive to light than L and M cones, blue light is not perceived very well under dim light conditions. As the intensity of a given type of light, such as incandescent light, increases, the perception of colors changes. A similar phenomenon occurs at the end of the day. As sunlight fades, the colors of the flowers in a garden, for example, will change in subtle ways. Artists such as Claude Monet explore this phenomenon in their paintings. Many of Monet's canvases convey a feeling for the time of day when they were painted.

### PHOTOTRANSDUCTION

Even though the retina is only 200 to 300 micrometers thick (eight to twelve thousandths of an inch, or two to three sheets of paper), there is much more there than just a layer of rods and cones. Depending on how you count them, there are as many as ten layers of cells in the retina. Light impinging on the retina has to pass through eight of those layers before finally arriving at the rods and cones. Any light that manages to get past the rods and cones is efficiently absorbed by the final layer, behind the rods and cones. This keeps light from bouncing around inside the retina, ensuring that the only light energizing the rods and cones is that entering through the front of the eye, not anything reflecting off the back of the eye.

But what about the other impossibly thin layers that lie on top of the rods and cones? Their job is to collect the electrical signals generated through chemical reactions in the rods and cones and send the signals on their way out of the eye, toward the brain. This process is complex in the extreme; although it has been intensively studied, it remains imperfectly understood. Let's begin with the role of the rods and cones, the transducers in the system.

Transduction is the process of converting one form of energy to another. For each of our senses, the energy contained in the stimulus is converted into electrical energy. The processes by which this occurs vary depending on which sense we're talking about. In vision, the energy contained in light is transformed through chemical reactions in the rods and cones into an electrical current. Strictly speaking, a photovoltaic solar cell does the same thing—it converts incoming light energy into outgoing electrical energy. But the physics and chemistry of the way a solar cell transduces light is totally different from the way the retina does it.

In the retina, a bewildering array of chemical-reaction cycles are set in motion both by light and by its absence. Which cycle occurs depends on how much light there is. Different reaction cycles operate in conditions of bright, medium, and low light levels, and the reactions that occur in the three types of cones differ. A completely different reaction cycle operates in conditions of profound darkness. This dark-adaptation cycle conditions the eye, slowly, for seeing in dim light.

The fuel for the chemical reactions in the retina is supplied by the body's normal metabolic processes, with one exception. A vital chemical called retinal is derived from food sources containing vitamin A. Vitamin A is famously present in carrots but also in lots of other orange or dark green plants, such as sweet potatoes, cantaloupe, kale, and broccoli. Very large concentrations are found in the livers of cows, pigs, and chickens. One medium-sized carrot provides the recommended daily allowance of vitamin A for an adult male. When I was a kid, jokes about carrots and vision were popular for a while. Mel Brooks told a story about a guy who ate so many carrots that he couldn't sleep at night because he could see through his eyelids. The reality isn't quite that dramatic, but it is well established that an adequate supply of vitamin A is essential to both color and night vision.

The only thing that light does in all of these chemical reactions is kick off the very first step in the reaction cycle. That first step is always the same: the transformation of retinal from one molecular form to another. That reaction sets off rapid cascades of other reactions that vary depending on the intensity and wavelength of the light and on which of the four types of receptors is involved. Under low light conditions, such as starlight, only rods operate. In somewhat brighter light, such as moonlight, both rods and cones operate together. When the light is brighter than that, our vision is dominated by the cones.

But these brief descriptions do not convey the remarkable ability of the retina to provide useful vision over a wide range of lighting conditions. The luminous intensity of light judged by the average person (in psychophysical tests, as described by Stevens) to be ideal for a task such as reading is about a million times more intense than the faintest light that person can perceive. And light can be a thousand times more intense than ideal reading light before the pain threshold is reached. Light so intense that it causes pain is thus about a thousand million, or $10^9$, times more intense than the faintest visible light.

Meticulously detailed studies of the electrical responses of single rods or cones to brief flashes of light of varying intensity and wavelength confirm that cones are much less sensitive than rods. Cones produce a voltage change smaller than that of rods, even when the cones are exposed to light thousands of times brighter. But these studies also show that, when exposed to a light flash, a cone's voltage spikes rapidly and returns to normal much more quickly than a rod's.

The chemistry that occurs in the three different types of cones is slightly different. Color vision depends on having different photoreceptors, the three types of cones and, to a lesser extent, the rods, optimized for light of different wavelengths. There are slight but crucial differences in the molecules to which retinal is bonded in the four types of photoreceptors.

In some ways, more goes on chemically in the absence of light, inside the retina, than when light shines. John Nolte notes that it might be more correct to refer to our rods and cones as "darkness receptors" than as light receptors. The complex chemical reactions that occur between a rod or a cone and the cells surrounding it in the retina speed up as illumination decreases. Light impinging on a rod or a cone causes reactions inside it that actually close a gate in the cell, slowing or stopping the chemical exchange between the cell and its surroundings. Closing that gate—stopping all that chemistry—is what causes the cell to polarize, or change its voltage.

The voltage changes in individual rod and cone cells are the inputs to the brain's perception system. Somehow, the brain manages to collect, organize, and interpret countless electrical spikes just a few thousandths of a second long, which come from tens of millions of photoreceptor cells. The end result is our visual world.

### JAVAL AND SACCADES

The fovea, at the center of the retina and of our visual field, does not focus on any one location for very long. Tiny, rapid movements of the eye muscles serve to constantly refocus the fovea. These movements are so rapid that the eye is constantly vibrating between 30 and 70 hertz (cycles per second). Each individual movement serves to reposition the fovea by one-third of a degree of the field of vision. The fovea thus flits about, never resting its gaze on anything for more than a fraction of a second. All this frenzied activity is completely transparent to us. We have no idea that we're doing it, or that anyone else is.

Better known than these tiny movements are the much larger jumps or saccades that our eyes perform when, for example, we are reading. Saccadic eye movements were first officially recognized and recorded in the 1880s, by the French ophthalmologist Emile Javal (1839–1907) and his co-workers. Javal became interested in ophthalmology because his younger sister, Sophie, was severely cross-eyed; she had a medical condition formally known as strabismus. Javal wrote his doctoral dissertation on strabismus. A system of exercises that he developed to retrain the eyes of strabismus sufferers proved effective for many of those afflicted, including Sophie.

In 1884, Javal was commissioned to study the visual aspects of the reading process, and it was there that he discovered and chronicled what are now called saccadic eye movements. He employed the French word *saccade,* which means a quick, irregular movement, to describe his observations of the eye, and the word came into English with this specific optical meaning. Javal concluded that a saccade, when a person is reading, is about ten letters long.

Utilizing their prodigious powers of observation and some clever experiments, Javal and his associates were able to deduce and quantify the saccadic nature of eye movements during reading. One critical experiment involved an ingenious setup wherein the saccadic motion of the eye during reading was transformed into sound, using a tiny flexible rod attached to the upper eyelid of the subject. Each saccade caused a tiny movement of the rod. The rod was bonded to the center of a taut membrane. When the eye moved, the rod moved, causing the membrane to vibrate, an arrangement that may well have been inspired by the eardrum and the tiny bones attached to it. Each eye movement transmitted a brief burst of sound to a microphone. The noises produced by saccadic motion were distinguishable from other, larger, eye movements, such as when the reader's focus moved from the end of one line to the beginning of the next.

One of Javal's technicians, a M. Lamare (the M. is for "monsieur"), even claimed to be able to deduce saccadic motions of the eye by resting his fingertips lightly on the upper eyelid. I've tried it, and while it's easy to detect major motions of my eyeballs in this way, discerning the much finer saccadic movements is a different matter. A lighter touch and a more practiced hand are clearly required. Javal and his team devoted their careers to investigations like this, which led to such fundamental discoveries as saccadic motion.

In a bizarre twist of fate, Javal developed glaucoma in middle age and was profoundly blind by 1900. The last years of his life were given to research on blindness, culminating in a book of practical advice for the blind and a machine to aid the blind in handwriting.

## The Evolution of the Eye

We saw in part 1 that eyes developed the ability to sense radiation in the 400–700 nm range because such light is not attenuated very much as it passes through water. Since the earliest seeing creatures lived underwater, it is logical that vision would evolve to sense electromagnetic waves

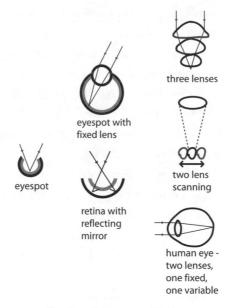

*Fig. 9.* Different ways of seeing things: a rudimentary eyespot, as in green algae; an eyespot to which a spherical, fixed lens has been added, found in some mollusks; a retina with a reflecting mirror behind it to increase illumination, as in the scallop; the three-lens arrangement of *Pontella;* the telescope-like two-lens eye of *Copilia,* in which the bottom lens scans from left to right; and the human eye with its arrangement of two lenses, one fixed and one variable. (Redrawn and adapted from Land and Fernald, "The Evolution of Eyes")

in the range we now call visible light. But this is only the beginning of the story of the evolution of the eye.

After Darwin, the evolution of the eye has been extensively studied. Eye structures are believed to have evolved in two major steps. The first was the development of "eyespots," which are visible-light analogues to the IR pit organs in the snakes discussed earlier. Eyespots, shallow cups containing a relatively few receptors sensitive to visible light, are still found in many species, but they are useful only for distinguishing light from dark, not for the detection of images or patterns. The other major stage in the evolution of the eye was the development of some sort of optical system that could allow photoreceptors to do more than just detect light.

The number of distinct kinds of optical systems that have evolved is remarkable. Some of the various functional developments, beginning with the eyespot, are shown in figure 9. Previously unknown optical arrangements in the animal kingdom, such as the telephoto lens of the chameleon, first discovered in 1995, continue to be revealed.

## Color Vision

Monochromatic (black and white) photography preceded color photography, and the same was true for television. It is believed that color vision also evolved much later than monochromatic vision. Color vision was an important development, endowing species with advantages in terms of such biological imperatives as finding food, detecting enemies, and locating suitable mates.

Certain bacteria possess a rudimentary ability to detect color. Visible light above a certain wavelength attracts the bacteria, while light below that wavelength repels them. These bacteria possess the two characteristics that any color vision system must have—at least two different mechanisms for absorbing light, each optimized for a different wavelength, and the ability to keep track of how much light each is absorbing.

Humans and other primates have evolved a system of color photoreceptors of intermediate complexity compared to other animals. With three types of color photoreceptors, humans have fewer than chickens (which have five) and more than mice (which have two). Human color photoreceptors, the three types of cones, are optimized for light of 440 nm, 540 nm, and 580 nm. Until about 30 or 40 million years ago, it is believed, only a single photoreceptor of wavelength greater than 500 nm existed in primates. The so-called red-green split that resulted in the presence of both the 540 nm and 580 nm cones has been the focus of much study by geneticists and vision experts.[2]

To explain how having three types of cones enables color vision, let's start with a visit to my tomato garden. We're having a particularly hot and wet summer this year, and my garden loves it. It's not so much a garden as a jungle. The twenty or so tomato plants were over six feet tall by late June, and they've grown together and fallen on one another in ways that might lead one to believe no one is taking care of them. Perish the thought. I water and weed my garden, and every day I paw through the tangle of vines and leaves, searching for reddening fruit in a sea of green. And every day I'm thankful I'm not red or green colorblind, because if I were, I wouldn't have a clue which tomatoes to harvest.

In individuals with normal color vision, the distribution of the three types of cones shows surprising variation. On average, the S cones account for only about 5 percent of the overall cone population. The relative amounts of L and M cones vary wildly, from about one to one to more than fifteen times as many L than M cones.[3] The genes for the L and M cones, having only "recently" (30–40 million years ago) separated, are

[2]The three human cones are sometimes misleadingly referred to as blue (440 nm), green (540 nm), and red (580 nm). The "red" cone is optimized for 580 nm light, which is actually orange-yellow. It is true that 580 nm cones are extremely important in distinguishing color at the red end of the visual spectrum. But the more widely accepted names for the three types of cones are S (short, 440 nm), M (medium, 540 nm), and L (long, 580 nm).

[3]Evidence of this variation comes from remarkable images made of the retinas of living subjects, using advanced optical techniques, as reported by H. Hofer and colleagues.

located next to each other on the X chromosome, and birth defects involving these two genes are relatively common. About 2 percent of males are either red or green colorblind. Those lacking L (red) cones suffer from protanopia, while those missing the M (green) cones have deuteranopia. Missing the gene for S (blue) cones is called tritanopia, but this is extremely rare. In each of these three forms of color blindness, the sufferer possesses only two types of cones instead of three. Such individuals possess "color vision" but of a vastly inferior sort. It has been known for centuries that some men have great difficulty harvesting fruits and vegetables based on their color. We now know why.

At first glance, it seems unlikely that the M and L cones, with their optimum wavelengths so close together (540 nm and 580 nm), would add so much richness to our color vision. A fascinating report by Jeremy Nathans of Johns Hopkins University sheds light on the matter. Nathans analyzed normal and colorblind eyes both experimentally and mathematically, in order to determine the ability of these eyes to discriminate colors of various wavelengths. A normal eye, with S, M, and L cones optimized for light at 440, 540, and 580 nm, is least able to discriminate colors with wavelengths of 440 (blue-violet) and 560 nm (yellow-green). But even so, in general, a normal eye does a pretty good job of discriminating colors all the way across the visible spectrum from 400 to 700 nm. A colorblind eye, one that is lacking either L or M cones, is a completely different story. Such an eye is virtually incapable of discriminating colors at 440 and 560 nm.

## Bandwidth of Vision

The personal computer and the digital video camera have caused us to think somewhat differently about how information is passed from the various sense organs on to the brain. In an experiment reminiscent of Descartes's work with the eyes of oxen, a rather gruesome exhibition in which he first demonstrated that the lens of the eye inverts images, researchers at the University of Pennsylvania wired a guinea pig retina to an array of electrodes and then exposed the retina to various visual stimuli. They concluded that a guinea pig eye transmits data to the brain at about 0.9 megabyte per second. Extrapolating to the larger size of a human retina, with about ten times as many cells for transmitting information from retina to brain, the researchers concluded that the human eye transmits visual data at about 9 megabytes per second, or about the rate of a high-speed Internet connection in 2009. It further appears, based on their research, that the eye is capable of much higher rates of data transmission than this—perhaps up to several hundred times faster.

If much higher rates of data transmission are possible, then why aren't they used? One explanation is the energy costs involved. The brain accounts for only about 2 percent of body weight, yet it consumes about 20 percent of all the energy the body burns. The brain thus consumes energy at ten times the rate, per pound, of the body's average. The brain would consume even more energy if the eyes transmitted data at higher rates.

## Visual Acuity

Do you have twenty-twenty vision? I wish I did. Someone with twenty-twenty vision is said to be able to see, at a distance of twenty feet, what a person with normal vision can see at twenty feet. A person with twenty–one hundred vision can see at twenty feet what someone with normal vision can see at one hundred feet. Eye care professionals tend to avoid describing vision with terms such as twenty-twenty; there are more precise and more useful ways to quantify our visual capabilities. For example, I am nearsighted, or myopic. This is a rather common condition; about one-third of all people are myopic to some extent. Myopic people can focus on things that are close up but not far away. I am pretty darn myopic. I can see things clearly, without glasses, only when those things are within about four or five inches of my eyes. From the very first time I had my eyes checked, at about age five, I have been unable, without glasses, to read the big "E" on the top row of the eye chart.

Eyeglasses can correct myopia. The light-bending ability of an eye-glass lens is measured in diopters (*diopter* means "to see through," from the Latin *dioptra*). A hand-held magnifying lens can bend sunlight rays, focusing them so intensely on a pile of dry leaves that the leaves might even catch fire. A 1-diopter lens takes parallel light rays and bends them so that they focus at a point 1 meter behind the lens. A 2-diopter lens focuses the light at a 0.5-meter distance, 4 diopters at a 0.25-meter distance, and so on. The diopter rating of a lens is thus equal to the inverse of its focal length. Lenses that have a negative diopter rating cause light rays to diverge instead of focusing them. Diverging lenses, with negative diopter measurements (mine are about minus 11) are used to correct myopia. In a myopic eye the light rays are focused in front of the retina, near the back of the eye, instead of right on the surface of the retina. A negative-diopter lens corrects this problem by causing the light rays to diverge just enough that they focus on the retina.

Eyeglasses have existed since at least 1300. Contact lenses were described and sketched by Leonardo da Vinci in 1508 but did not become practical until about 1940. Laser surgery to correct myopia and other vision defects began in the middle 1980s and was first approved in the United States in 1995. Such surgery involves reshaping the cornea at the front of the eye to modify its light-bending abilities. The cornea is like a natural contact lens. When it is surgically reshaped, it bends light differently than it did before. It's like getting a contact lens with a different prescription.

There are those who believe there is a downside to correcting myopia with ever-stronger negative-diopter lenses. These folks see evidence of a vicious circle—strong lenses that correct a certain level of myopia lead to an increase in myopia, and thus to even stronger lenses, and so on. It is true that those of us who are highly myopic are at greater risk than the general population for a detached retina (a frightening but generally surgically correctible condition) in later life. The link between such problems and the wearing of progressively stronger eyeglasses is somewhat tenuous.

## Keep Your Eye on the Ball

There is an adage in sports that the first things to go in an aging athlete are the legs. Pat Leahy, a place kicker for the New York Jets, disagreed. "It's the hair," he said in 1991. It might be neither. It could be that the eyes are the first to go, depending on which sport you play.

Among professional athletes, golfers tend to enjoy some of the longest careers. Jack Nicklaus claimed the last of his record eighteen major championships at age forty-six when he won the Masters in 1986. The Golden Bear played his final Masters in 2005 at sixty-five. In the summer of 2009, a fifty-nine-year-old Tom Watson elevated baby boomer golf fans to near ecstasy as he came within a single stroke of capturing the British Open. But a ten-foot putt for the win wouldn't fall, and he lost in a playoff.

Top athletes are forced to retire when they can no longer successfully command their bodies to perform at the highest levels. A wide receiver in football might "lose a step" and thus no longer be able to elude defenders the way he once did. An aging baseball shortstop faces a similar dilemma in trying to chase down hard-hit ground balls.

But golfers don't have to run fast, so what is it that finally forces them to hang up their spikes? It might be their eyes. No less an authority than Nicklaus complained that a loss of visual acuity late in his career was a significant factor in the deterioration of his golfing abilities. At the highest level, golfers require exceptionally sharp vision on the putting green, and putting has always separated the very best golfers from the rest. "Drive for show, putt for dough," the saying goes. To be a great putter, a golfer has to be able to see the tiniest nuances in the slope and changes in the surface of the green, all the way down to a single blade of grass bent the wrong way. Ignorance of these details means a missed putt, and a single missed putt often means a lost championship.

Vision is so important to top golfers that Tiger Woods, Vijay Singh, Zach Johnson, Fred Funk, and many others have undergone laser eye surgery to improve their vision to twenty-fifteen or better, in order, primarily, to improve their putting ability.

Efforts to improve vision among top athletes are not limited to eye surgery. There is the legendary story of 2005 U.S. Open golf champion Michael Campbell, who performed a strict regimen of eye exercises designed to improve his ability to see the proper line of his putts. So strict was it that he made frequent stops at portable toilets in between holes at the U.S. Open in order to repeat an exercise in which he waved a pencil in a horizontal figure-eight pattern in front of his face for about two minutes, all the while following it with his eyes. Does this sort of thing work? Well, Campbell did win the U.S. Open.

Golfers have no monopoly on the need for exceptional visual acuity. Hitting a baseball is often cited as one of the most difficult tasks in sports, and razor-sharp vision is among the prerequisites for success. Joe DiMaggio credited his twenty-ten vision for some of his legendary ability. DiMaggio's contemporary Ted Williams had similarly sharp vision; it was said that he could read the label of a record while it was spinning on a record player, a parlor trick one rarely gets the chance to see repeated in the age of the iPod. Pete Rose speaks eloquently about the different visual appearances of various types of baseball pitches, from the batter's perspective. A slider appears to the sharp-eyed batter as a white circle with a red dot inside it. When the pitcher throws a slider, the red stitches on the white ball spin in such a way that the batter, if his eyes are good enough, sees a red dot on the ball. The better his vision, the sooner the batter picks up the red dot, and the more time he has to react and hit the ball.

Wide receiver Larry Fitzgerald thrilled the football world when he led his Arizona Cardinals to the 2009 Super Bowl. Among Fitzgerald's many athletic gifts is his vision, including a renowned ability to pick up the flight of the ball at the very last instant. At the highest levels of the game, football receivers often do not have the luxury of seeing the ball leave the quarterback's hand and arc its way across the field. Having run his route, at the last instant the receiver turns and looks to where the ball should be. In a tiny fraction of a second, he must pick up the flight of the ball and adjust to it in time to catch it. Fitzgerald is a master at this, and he credits a lifetime of eye exercises. As a youngster, for example, he would play catch with one eye or the other closed, catching the ball one-handed with the hand on the side of his closed eye.

Over the past twenty years, eye exercises for athletes and others have gained popularity. The goal of these exercises is to train and fine-tune the muscles that control the movements of the eyes. The benefits for athletes are promoted in terms any athlete can relate to. Athletes spend so much time training their muscles. So why neglect a set of muscles so intimately related to the most important sense in almost any sport—your vision? When your sport requires you, for example, to track the flight of a ball that moves at upwards of 100 miles per hour (a pitched baseball), 150 miles an hour (a served tennis ball), or 200 miles an hour (a smashed badminton shuttlecock), it makes perfect sense to try to improve the performance of the muscles that control the movements of your eyes.

# Chapter 6 • The Chemical Senses

Aristotle's neat classification of five senses gets us into trouble lots of ways, but perhaps nowhere does it create more problems than when we try to understand how we sense what's good and bad about food and drink.

Everyone is aware that the tongue is a sense organ. It contains receptors that allow us, in everyday language, to "taste" food. What the tongue does is technically known as gustation. It's also relatively common knowledge that the tongue does not act alone when it comes to sensing flavor. The nose has a strong role to play as well; the technical name for what the nose does is olfaction. We both smell and taste the things we eat and drink. But that's hardly the end of the story. There is also the common chemical sense, which is attributed to various nerve endings in the mucous membranes of the mouth and nose that are responsible for sensations such as the hotness of chili peppers, the burning sensation of ammonia fumes, and the cool flavor of menthol. Still other important aspects of flavor include the texture and temperature of food. Touch and temperature are not chemical senses, though, and we'll save our discussion of them for chapter 7. For now, we'll focus on the chemical senses, primarily smell and taste, in which sensory receptors are reacting to the chemical properties of molecules.

And then there is the special case of the pheromones, another chemical sense. The pheromone response is important for many animals, but its role in human beings is not well understood and is somewhat controversial.

Pheromones are chemicals produced by some plants and animals that evoke specific responses, usually related to reproduction, in other members of the same species.

Organisms that utilize the pheromone response include insects, rats, reptiles, cats, and even some plants. Animals that sense pheromones usually possess a specialized organ, located adjacent to the bone that separates the two nostrils. This organ contains receptors that are similar in many ways to the smell receptors in these animals.

The first pheromone to be discovered was an alcohol called bombykol, produced by female moths as a sex attractant. Male moths can detect bombykol in vanishingly small concentrations. The human nose is far less sensitive. The moth is about one hundred thousand times more sensitive to bombykol than the human nose is to any known chemical. Once a moth has detected bombykol, it has the ability to fly in the direction of increasing concentrations of the chemical, and at the end of the voyage, it will have located a potential mate.

Human fetuses possess pheromone organs, but they degenerate and disappear during childhood. Pheromonal activity in humans, if such a thing exists, would take place through the sense of smell. What might those pheromonal activities be? Perhaps the best-known and most-studied effect potentially related to human pheromones is the menstrual cycles of women living in the same house. There has long been anecdotal evidence that women living together tend to begin menstruating at the same time each month. Scientific studies on this subject have had varying results. In 1971, the psychologist Martha McClintock published the results of a study in *Nature* that purported to show that women living in close proximity do indeed begin to menstruate at around the same time. What has been termed the "McClintock Effect" has been studied by others, and the effect has not always been confirmed. Such studies are complicated by lots of factors, including the fact that women's menstrual cycles are not of the same length.

In any event, there remains a lot of interest in the potential existence of pheromonal activity in humans. Not all of that interest is purely scientific. The fragrance industries, for example, would be particularly keen to add a dash of an odorless chemical with the potential to sexually attract men, women, or both. The applications of pheromones in perfume and cologne are obvious—and perhaps a little frightening. There would also be plenty of interest from manufacturers of shampoo, soap, laundry detergent, and other products.

# What the Nose Knows

Stephen D., a young medical student, experienced a vivid dream one night that was probably induced by his abuse of cocaine and other drugs. "I had dreamt I was a dog—it was an olfactory dream—and now I awoke to an infinitely redolent world—a world in which all other sensations, enhanced as they were, paled before smell." For the next three weeks, Stephen's sense of smell was hypersensitive, and it took on an exaggerated importance in his life. When he went into his clinic at the hospital, he recognized each patient's smell before he saw him or her. He found this "smell-face" more vivid and suggestive than the visual face he had become used to. Smell dominated his world. He found he could navigate his New York City neighborhood by smell alone.

During this period, things didn't seem real to Stephen until he could smell them. This formerly reflective, abstract intellectual now lived in a world of "immediate significance," in which intellect mattered little. But as quickly as it had begun, the sensory transformation ended. His sense of smell returned to normal, and he went back to a life of abstract reflection and relatively faint senses. Years passed, and the incident never repeated itself. But Stephen, now a successful doctor, occasionally yearned to revisit this other world. "I see now what we give up in being civilized and human. We need the other—the 'primitive'—as well. . . . If only I could go back sometimes and be a dog again!" Stephen's story is "The Dog Beneath the Skin," in Oliver Sacks's *The Man Who Mistook His Wife for a Hat*. We will meet several other characters from that book later.

Smell is perhaps the most mysterious of the senses. The mysteries encompass aspects of stimulus, sensation, and perception. Although there are several strong theories, we still don't know what it is about airborne molecules that make them smell different from one another. We understand at a technical level, in terms of how the hardware works, why our eyes see a rose differently from a violet and why a violin sounds different from a clarinet, but we do not know exactly what the olfactory receptors inside our nose sense to let us know that it's an apple pie we smell in the kitchen, and not a rotten egg.

Writing in the *New England Journal of Medicine* in 1980, Lewis Thomas contemplated the search for an understanding of the sense of smell. He noted that overall advancements in the biological sciences might be measured, centuries from now, by how long it will take to gain "a complete, comprehensive understanding of odor. It may not seem a profound enough problem to dominate all the life sciences, but it contains, piece by piece, all the mysteries."

Progress has been made since Thomas wrote those words in 1980. Richard Axel and Linda Buck received the Nobel Prize for Physiology or Medicine in 2004 for their research on "odorant receptors and the organization of the olfactory system." Much of what we know about the apparatus of the sense of smell has come from the research they reported in 1991. Why did so much time pass before humankind was able to sort out some of the basics of how the hardware of the sense of smell functions? By 1991, many of the mysteries of the other senses had long been unraveled. Georg von Békésy's Nobel Prize for research on the inner ear was awarded in 1961, and a Nobel Prize for vision research goes all the way back to 1910. But one might just as well ask why, even after the pioneering work of Axel and Buck, there remain so many fundamental questions about the sense of smell.

The sense of smell is an instrument for identifying airborne molecules. Even single-celled organisms possess some sort of analogous sense with which they evaluate their chemical surroundings. Lacking such a sense would render nearly any species nonviable. Animals, humans included, use olfaction to identify food sources, detect predators, and find mates.

Not all molecules can be identified by the nose. Oxygen, nitrogen, and methane, for example, are considered odorless. In the case of oxygen and nitrogen, this is probably a good thing, given their omnipresence. Oxygen makes up about 21 percent of air, while nitrogen is the most common element at 78 percent. Methane is the predominant molecule in natural gas, and its lack of odor can be a problem, since it is piped into many of our houses. Natural gas leaks can be quite dangerous. To help our noses detect such leaks, tiny doses of powerful odorants are added to natural gas. Often, these additives are sulfur-based mercaptans. I investigated an accident in which a stainless steel container containing several hundred gallons of mercaptans had developed a leak. The mercaptans had been traveling in a container stored in the trailer of an eighteen-wheeler on the highway. Pretty soon after the container began leaking, the smell overwhelmed the driver, and he was forced to pull off the highway and abandon his vehicle. The leak was relatively small, and most of the product remained inside the container. Nonetheless, even after the guys in the hazmat suits had finished cleaning up the truck so that we could investigate, the mercaptan smell was almost more than we could stand.

Even if we can't detect methane, we can detect a staggering number of other molecules, up to about ten thousand, with our noses. Single-molecule odors, such as those of acetone or hydrogen sulfide, are easily identified. But so are complex blends of many chemicals, including the aroma of a fresh peach, a glass of chardonnay, and a Cuban cigar.

Part of Buck and Axel's contribution to our understanding of olfaction was their discovery of a startlingly large collection of genes, initially found in rats and later in humans and other animals, whose purpose is to encode, and thus create, smell receptors. There are well over 1,000 such genes in rats, and 350 or more in humans. Of the 25,000 or so genes in the human genome, more than 1 percent are devoted to the sense of smell. Some put the percentage as high as 3 percent. When one contemplates all the different functions of the human body and all the diverse, highly specialized cells that exist to carry out those functions, it is astonishing to think that so many genes are devoted to the sense of smell.

## *The Psychology of Smell*

As well, it gives one pause to realize what ancient creatures we really are. It may indicate to what extent we are ignorant, in our everyday consciousness, of the true nature of the beings that we have evolved into. At some point in our history, all those genes that create olfactory receptors were evolving, enabling us to recognize an almost unimaginably wide spectrum of chemicals through our noses. This fact must have profound influences, both conscious and unconscious, on our existence. We can smell all that stuff. There's got to be a reason for it.

The psychological aspects of olfaction were considered by Plato. He generally downplayed the importance of the sense of smell, declaring that vision and hearing were more important and nobler senses than either smell or touch. Smell has likewise received relatively short shrift from other philosophers throughout most of recorded history. When it has been considered, the attention has often been negative. Smell was the sense of lust and impulsive behavior. Vision and hearing were the sources of knowledge and reason.

Smell is a visceral sense, with strong ties to our subconscious, and as such it did not escape the notice of Sigmund Freud. Freud, among others, noted that because human beings walk upright, our eyes have the high ground and they often allow us to see things well before we can smell them. Dogs get around on all fours, and their long snouts suggest exactly what they are all about sensually: smell. With their nose to the ground, where odors, many of them heavier than air, linger, the dog's sensual world could scarcely be more different from our own, as anyone who has ever taken a dog for a walk knows. But, as Oliver Sacks reminds us, there is a dog beneath the skin of every human being. The creatures we have evolved from did not always walk upright, and there was a time when their sensual world was probably not all that different from that of today's dog.

Smell has been called the most ancient of the senses. It is known that the parts of the brain in which olfactory sensations are processed are generally the parts that evolved early on. Maybe as the other senses, especially vision and hearing, evolved, smell became progressively less important. Nonetheless, the links between the olfactory receptors and parts of the brain that have to do with our emotions and motivations ensure that the sense of smell retains a vital role in determining our nature and shaping our behavior.

## How Olfaction Happens

Much as the inner ear is located well inside the skull, downstream of the external ear, so the smell receptors are positioned downstream of the external nose. Airborne molecules arrive from two directions: the front door is the nose, and the back door the retronasal passage, an airway that connects the olfactory center to the back of the mouth. We continue to smell food even after it is entirely inside our mouth and while our mouth is closed.

Each smell receptor is able to recognize multiple odors, and individual odors can activate multiple receptors. This helps account for the very large number of odors, up to about 10,000, that most of us can distinguish, since there are only about 350 different types of smell receptors in the nose. On average there are roughly 30 different recognizable odors for each type of receptor. The idea that once held sway was that a one-to-one correspondence existed between odors and receptors. But it is now recognized that each type of olfactory receptor can be stimulated by quite a few different kinds of odorant molecules and that there is a lot of overlap among the different receptors in terms of which molecules they are stimulated by. Exactly how a molecule activates an olfactory receptor is far from clear, although there are several prominent theories. Recall our discussions of shape versus frequency in part 1.

Contrast the large number of different olfactory receptors with the four different types of receptors in the retina, which allow us to distinguish millions of different colors. Why does smell require so many receptors?

Even if the exact mechanism of an individual smell receptor were understood, there would remain more than enough olfactory mysteries. For example, it is not well understood how the olfactory nerves, to which the olfactory receptors are connected, can pass along any sort of categorization of odors, since the contribution of each olfactory receptor is intimately blended with those of the others. A photographic analogy is sometimes used here. Individual olfactory nerves might be compared to pixels in a digital photo. Each provides a small part of the overall image, or in this case one characteristic of the overall smell. Those "pixels" must somehow be assembled into that overall smell, a process that seems to take place inside the olfactory bulb of the brain. And here's one more mystery: Because we lack a comprehensive theory of the stimulus of olfaction, we can neither predict what a certain molecule will smell like nor even whether we will be able to smell it at all. Apart from the purely scientific question of the basic mechanism of olfaction, who really cares? Well, there is a lot of money to be made from the development of new odors for products such as perfumes and detergents, as we described in chapter 2. The mysteries surrounding the sense of smell make such work much more difficult.

The smell receptors are found in a patch of tissue about half the size of a postage stamp on the roof of the nasal cavity. Odor molecules must travel from the entrance to the nostrils about 7 centimeters (2.75 inches), along a path that arcs upward toward the back of the skull, to a location roughly level with the eyes and behind them. The 3 million or so receptor cells found there are to the sense of smell what the retina is to vision or the organ of Corti is to hearing.

Under a microscope, the surface of the receptor cells reminds me of a living coral reef. An individual smell receptor cell consists of tentacle-like growths, called cilia, emerging from the stem of the cell underneath. The cilia are analogous to the rods and cones in the retina. When an individual cilium is stimulated by any one of the odorants to which it is sensitive, a cycle of chemical reactions is set into motion that results in a voltage change in the receptor cell. As with vision in the retina, this voltage change is the all-important beginning of olfaction.

The three types of cones in the retina are optimized for short-, medium-, and long-wavelength light, but there is a lot of overlap in the wavelengths each type of cone absorbs. So it is with the olfactory receptors. Each type of receptor absorbs stimuli from an array of odorants. The odorants that any one type of receptor is stimulated by are different from, but overlap with, those that stimulate other types of receptors. The roughly 350 different types of olfactory receptors present quite a contrast to the 4 types of receptors utilized for vision. Once a given olfactory receptor has been stimulated, the resulting electrical signal begins a complex journey to the olfactory regions of the brain, where an odor is perceived.

## Smell Disorders

In the world of medicine, the area of smell disorders is a prime example of a relatively wide-open field. The number of researchers working in this area is tiny compared to the number working on vision or hearing disorders.

Perhaps as a result, disorders of the sense of smell, including its profound absence, were once believed to be relatively rare. In fact, they affect millions of people in America alone. Anosmia is the complete absence of the sense of smell, and as many as 2 million Americans may suffer from it.[1] That's 1 American for every 150. Just as a loss of vision with age is called presbyopia, presbyosmia is an age-related decline in the sense of smell. An overly sensitive sense of smell is called hyperosmia, and a partial loss of the sense is hyposmia. A person afflicted with parosmia has a distorted sense of smell, sensing as unpleasant odors what people with a normal sense of smell consider pleasant. Phantosmia is a condition in which the sufferer smells things, typically unpleasant odors, when no external stimuli are present. Some of these disorders are related to problems with the sensory receptors in the nose, whereas others are more perceptual—problems that originate somewhere in the brain.

Many blind persons use canes and specially trained dogs to get around. They often learn to read Braille. The deaf learn to read lips and communicate

---

[1] In comparison, approximately 1.3 million Americans are legally blind.

with sign language. How do people with smell disorders compensate? It seems odd that humankind has learned to train dogs to aid the blind. As anyone with a canine companion knows, what dogs are really good at is smelling things. Why aren't there smelling-nose dogs, to go along with their seeing-eye companions? Given the prodigious accomplishments of bomb-sniffing and drug-sniffing dogs, it would seem a natural fit.

Where, in the pantheon of human suffering, should we place those with disorders of the sense of smell? Are they, like the blind or deaf, persons with disabilities, or are olfactory disorders merely a nuisance, like baldness? For a long time, olfactory disorders were largely ignored by the medical community. This may have been because, for the modern human being, the sense of smell is not considered essential, as sight and hearing are. But more and more attention is being paid to olfactory disorders these days. For some people (cooks come to mind) the sense of smell is hardly a luxury.

## A Dog's Life

When I take my dog for a walk, her nose works overtime, taking in all the interesting smells at ground level in our neighborhood. Is she detecting the scents of other dogs? Other animals? People? Since humans are oblivious to practically all of these scents, it's difficult to know just what sort of data our dogs are processing, as they go about the serious business of nonstop sniffing. The information a dog gleans by sniffing a fire hydrant must be as different, from one day to the next, as the headlines in the newspaper are to me. What makes a dog's nose so much more sensitive than ours?

Quite a lot of things. With a long snout within easy reach of the ground, dogs place their nostrils right on top of what they want to smell, thus bringing in a higher percentage of odorant relative to other, less interesting molecules. Pound for pound, dogs inhale a larger volume of air through their noses than humans do, another way they bring more odorant to bear on their smell receptors. Furthermore, dogs have upwards of 200 million smell receptors, compared to the paltry 3 million allotted to humans. It takes a lot of computing power to process all that olfactory data, and up to one-third of a dog's brain is devoted to olfactory perception, far greater than the corresponding volume in the human brain.

These olfactory advantages combine to make dogs supercharged sniffing machines. Dogs have been domesticated for many thousands of years and have become exceptionally trainable. They are thus just about perfect for sniffing out bombs in war zones and airports and drugs at border crossings, airports, and crime scenes. One military dog in Afghanistan had sniffed out more than two thousand pounds of explosives and was still going strong. Dogs' olfactory abilities have long been utilized in the search for missing persons. And relatively recently, dogs have even provided evidence in courts of law through what are known as "dog-scent lineups."

In a dog-scent lineup, a trained dog is first exposed to the scents of items found at a crime scene. The dog is then walked past the olfactory equivalent of a police lineup: a series of containers filled with samples swabbed from the suspect and from various other people. If the scents in any of the containers match what the dog smelled at the crime scene, the dog will alert its handler. Such evidence remains controversial—it has been accepted in some courts and rejected by others. Numerous lawsuits have been filed by individuals claiming false imprisonment based on dog-scent lineups. When a dog-scent lineup amounts to barking up the wrong tree, as it were, the fault is often human, and not canine. It's very easy to cross-contaminate the scents, and dog-scent lineup procedures are sometimes not very well controlled.

## The Electronic Nose

While the future of the dog-scent lineup is unclear, the use of dogs for locating bombs, drugs, and missing persons remains quite popular. Nonetheless, the quest for an artificial sniffing machine to replace the dog continues. It takes a lot of time and money to train a dog to sniff out drugs and bombs, and the career of such a dog is relatively short—only a few years. In addition, even though dogs are hard workers, being made of flesh and blood, they cannot work nonstop. The hope is that a sniffing machine, in addition to saving money and time, might someday be even more sensitive, faster, and more reliable than a dog's nose.

The three requirements for excellent artificial olfaction, or what is called the "electronic nose," are the same features that give a dog its olfactory advantages. The first of these is the front end, or the sniffer—the hardware that draws in scent-laden air. Next are the sensors that detect airborne molecules from drugs and explosives. Finally, there is the computer that interprets the data and controls the system. Most of the research into artificial olfaction has focused on sensor technology. Sensors for detecting various odorants utilize a wide variety of scientific principles. For example, carbon monoxide detectors are cheap, readily available, and found in millions of homes. Their job is to detect this colorless, odorless, poisonous gas in indoor air. They are not "electronic noses" per se, but their sensors utilize some of the same technologies.

Three types of sensors have been used in commercial carbon monoxide detectors. These are metal oxide semiconductor, biomimetic, and electrochemical sensors. Metal oxide semiconductor sensors consist of thin wires of tin dioxide. The wires, when heated to about 400°C, are very sensitive to the presence of carbon monoxide (CO). When CO is present, the resistance in the wires decreases, causing an alarm to sound. This type of sensor consumes a lot of energy to heat the wires to 400°C, and they must typically be plugged in to household current, as opposed to operating from batteries. Biomimetic sensors employ a gel that changes color when exposed to CO. The color change is detected by an optical sensor, which sounds an alarm.

Electrochemical sensors have become much less expensive than they used to be. They utilize a form of fuel cell. Instead of hydrogen, the fuel in the fuel cells that power some prototype automobiles, these fuel cells consume carbon monoxide. As the concentration of carbon monoxide increases, more current flows through the tiny fuel cell in these sensors. When the current exceeds a certain level, an alarm sounds. Electrochemical CO detectors are accurate, consume very little energy, and have a long life. They have become the technology of choice in much of the world.

As artificial noses continue to improve, there may come a time when their abilities exceed those of their canine competitors. But how will we know it? Veterinary researchers at Auburn University have quantified the olfactory abilities of dogs for detecting various drug and bomb-related odorants. Most of the odorants tested reveal a classic "threshold" response. Below a threshold concentration, a given chemical is almost never detected by the dogs. Above the threshold, it is detected 90 percent of the time or more. Measured in these terms, electronic noses are getting very close to the sensitivity of trained dogs, in some cases even exceeding them. Such sensitivity studies do not tell the whole story, though, and electronic noses still require improvements in their sampling systems, their ability to over-come interference from masking odorants, and their limited mobility and robustness.

## The Sense of Smell in Medicine

Doctors put their senses to good use in diagnosing their patients' ills. It was Hippocrates himself who said, "In our art, there exists no certainty except in our sensations." And so doctors listen to our hearts and lungs through stethoscopes and look at our skin and peer into our eyes, ears, and mouths. They palpate, or feel with their fingers, various body parts, probing for tumors, fluid buildups, and joint problems. They smell us, too. It is well known that a dangerous, potentially fatal condition in diabetics known as diabetic ketoacidosis is generally preceded by a fruity odor in the breath of the patient. That odor is caused by a buildup of acetone (the main ingredient in nail polish remover) in the blood. The acetone is ex-pelled from the body via the bladder and the lungs, hence the symptom sometimes called "acetone breath." Any doctor or nurse with a properly functioning nose knows this odor, as well as what to do about it.

Scurvy, a form of malnutrition caused by insufficient vitamin C, re-sults in putrid breath. Mononucleosis is preceded by a sour breath odor. Various types of substance abuse can result in certain breath and body odors. And yet there is no serious, systematic use of olfaction in modern medicine. One reason is the lack of quantification in olfaction. Acetone breath in diabetics is only an indication. Diabetic ketoacidosis is easily verified through a quantitative urine test. The proliferation of highly sophisticated analytical techniques such as MRI has led to a de-emphasis on the use of the natural senses in medicine, including olfaction.

The sense of smell, however, has a long medical tradition. The idea that bad smells are unhealthy, still commonplace, is quite old.[2] Medicine's history with olfaction is rather checkered. The name for the disease malaria, for example, was coined in 1740 from the Italian *mala aria* and literally means "bad air." It wasn't until 1881 that it was first theorized, correctly, that malaria is transmitted not by the air but by mosquitoes.

As far back as the eleventh century, a comprehensive diagnostic system based on bodily smells, including breath, sweat, urine, feces, blood, and saliva, was developed. An early practitioner was the Arab physician Avicenna (980–1037). Historically, doctors seemed to be of two minds when it came to olfaction. On the one hand, they believed much could be learned about patients from the way they smelled. On the other hand, they were quite concerned about the bad things those smells might do to their own health. The stethoscope not only helps a doctor hear a patient's heart and lungs; it also allows him to do so while maintaining some distance from the source of a patient's body odors. Therapies involving scent, such as aromatherapy, herbal and mineral baths, and taking the "air cure" in the mountains have a very long history. In modern medicine in most Western cultures, however, such therapies are generally not considered "real" medicine.

Given the pace of improvements in the electronic nose, olfaction in medicine may someday come full circle. As we get better and better at artificially sniffing trace odorants in breath and other body odors, the quantitative

[2]The "badness" of a smell is at least partially a cultural phenomenon. Some smells are revered in one culture but reviled in another.

detection of various conditions by olfaction could become routine. Research into the use of artificial olfaction for the detection of staph, strep, E. coli, and urinary tract infections; for uremia and cirrhosis; and for tuberculosis and other diseases continues. These efforts face many challenges, above and beyond those related to the electronic nose. Some of the more daunting challenges are medical in nature. People are different. They don't always emit the same odorants, at the same levels, for the same diseases. Misdiagnoses, both false positives and false negatives, are a big problem. Finally, research funding in medicine tends to favor therapeutic investigations—the search for the magic pill—rather than diagnostic inquiries.

## Taste

The great writer and philosopher François-Marie Arouet, better known as Voltaire, was said to enjoy forty cups of coffee a day. He is widely accredited with having said, "Nothing would be more tiresome than eating and drinking if God had not made them a pleasure as well as a necessity." My father would be the first to admit that he is no Voltaire, although I would dearly love to listen to him discuss things over dinner with the celebrated Frenchman. I'll bet Dad would hang right in there. Next to a good conversation, there aren't too many things in life he finds more pleasurable than a good meal. Over eighty years old now, my father has always taken very good care of himself. He still walks two miles every day at a robust pace, and mentally he's very sharp. But because of Parkinson's disease, Dad lost the ability to swallow correctly, and his doctors were forced to surgically install a feeding tube a few years back. For nearly a year, Dad took nothing orally. All his food, water, and medicine arrived in his stomach not by way of the mouth and the esophagus, but through a plastic tube inserted straight through his belly.

This was quite a lifestyle change. It reinforced for me just how intertwined our social lives are with eating and drinking. Being around his family and friends while they were eating, drinking, and socializing was hard enough, but Dad told me one of the hardest things to get used to was the simple fact that taste was no longer a part of his life. He hadn't been deprived of the sense of taste. His taste buds still worked normally for a man his age. It's just that he didn't get the chance to use them anymore. They received no stimuli. It would be like sealing a sighted person into a profoundly darkened room, or perhaps sewing his eyelids shut. The eyes would still function, but there would be nothing to stimulate them. That's how it was with my dad and taste. All his nourishment bypassed his tongue and nose, entering his stomach utterly unsensed. He and my mother turned this into a little game. "That was a turkey sandwich," Mom would say, having injected yet another can of liquid nourishment. "Mmm," Dad would say. "Delicious."

Not long after his feeding tube was installed, Dad began a vigorous program of physical therapy designed to rehabilitate his swallowing muscles and reflexes. It's been pretty successful, and within a year he was getting more than half his calories the old-fashioned way—orally. I couldn't be any happier for him. Ever since my dad's feeding-tube operation, I find myself lingering over my meals a little longer and savoring their tastes a little more.

## The Elements of Flavor

The formal, rarely used term for the sense of taste is *gustation,* which is defined in the *American Heritage Dictionary* as "the act or faculty of tasting." There is a problem with the common name of this particular sense. *Taste* is not only "the sense that distinguishes the qualities of dissolved substances in contact with the taste buds on the tongue," but it is also "the faculty of discerning what is aesthetically excellent or appropriate." This double meaning has been exploited by innumerable advertisers and joke-tellers over the years. To wit, Charlie the Starfish tuna: "Charlie, Starfish doesn't want tuna with good taste, Starfish wants tuna that taste good."

Perhaps it is no accident that the English word we use to describe the actions of the sense organ in the tongue has another, broader meaning related to aesthetic matters in general. In French, *goût* has the same dual meaning, as does *gusto* in Spanish. The sense of taste is evidently so important that it has become synonymous, in multiple languages, with all things aesthetic.

Another peculiarity of the sense of taste is that it does not correspond uniquely with the perception of flavor in food. There are at least four different senses working together here. A little later I will describe a similarly multisensory phenomenon when we come to the senses that allow us to achieve the stability of our bodies, for example, an upright posture.

Strictly speaking, gustation refers to the sensations we derive from the taste buds on the tongue. A more general term is *flavor*, which integrates multiple senses into the overall experience we derive from food and drink. The sensations, the flavors, that we experience are nearly always a combination of taste (gustation), smell (olfaction), what is called the common chemical sense, and the senses of touch and temperature. The nose may be as important as the tongue in all of this, although the nose's contribution is difficult to measure.

But perhaps not impossible. I remember in junior high science class being fed small pieces of peeled raw potatoes and apples while blindfolded and with a clothespin holding my nose closed. Under those conditions, it was difficult to tell a potato from an apple. The tongue alone doesn't sense much difference between the two, and their textures are similar. With the clothespin limiting the influence of the nose, raw apples and potatoes aren't all that different. One problem with this little experiment is that there are *two* routes by which airborne molecules gain access to the olfactory sensors. In addition to the nostrils, there is also the retronasal passage at the back of the mouth. When food is chewed, any odors that are released can be sensed via this back-door passage, and thus smell is so intimately intertwined with taste that it is difficult to truly experience the latter in the complete absence of the former, except, perhaps, when you have a really bad cold. Individuals who complain that they've lost their sense of taste are sometimes discovered to have a disorder of the sense of smell instead.

The common chemical sense comes from various nerve endings in the mouth and nasal cavity that are responsible for sensing such aspects of flavor as the spiciness of peppers and the coolness of menthol. The texture and temperature of food are important too. These aspects of taste are perhaps underappreciated, but not in my household. My youngest stepdaughter is a wonderful kid, but she would be the first to admit that she has, in her own words, "texture issues" when it come to food. She will not drink any orange juice except the "pulp-free" kind. From the standpoints of taste and smell, I suspect the differences between regular and pulp-free orange juice are not significant. I conclude that what turns her off has to be related to texture and thus to the sense of touch. I'm sure she couldn't care less, as long as I buy pulp-free O.J. Whatever it is, this effect is powerful, and it isn't limited to orange juice. Other foods with lots of texture, things like strawberries, walnuts, and chunky peanut butter, also provoke her displeasure. Although she may be an extreme case, most people find texture to be an important aspect of flavor.

The influence of temperature on flavor also tends to be underestimated. The temperature of a food influences its flavor in terms of taste and smell, but much of the effect comes from our temperature-sensing ability. There are also cultural differences when it comes to food and temperature. The English seem to prefer toasted bread after it has cooled to room temperature, while most Americans want to consume it hot, straight out of the toaster. At the practical level, the Brits have got us Yankees licked. Toasted bread cools off very quickly, so you have to be quick to consume it hot.

Then there's the strange case of drinking water. In the United States, we nearly always take our water iced, a practice that is unusual elsewhere in the world. I was having lunch in the United States one day with a French friend who has lived here for many years. Joining us was a Frenchman on his first visit to the states. We'd just been seated when the waiter arrived with large glasses of water filled with ice cubes. Our visitor was puzzled by this. He asked us the reason for this iced water custom. Before I could say a word, my ex-patriot friend replied, "They put ice in the water over here because that way you don't taste the water."

I had never thought of it that way, but I eventually decided she was at least partially right. Chilling water to the freezing point does have the effect of subduing whatever flavors, pleasant or otherwise, the water may contain. Unless the local tap water is really foul, a restaurant can get away with serving just about anything, provided they serve it on ice. Water does have a taste that varies by region. The longest-running taste competition for tap water is the Berkeley Springs International Water Tasting, held in Berkeley Springs, West Virginia. There, they have been judging water quality in North America since 1991. The tap water of the aptly named community of Clearbrook, British Columbia, captured the gold medal in 2009.

I've discussed the ice-water theory of my French friend with lots of American friends. Often, they find the taste explanation for why we ice the water to be outrageous, even bordering on offensive. They tell me my French friend and I are nuts, or worse. We put ice in water in America, they say, because it helps cool you down on a hot day. I used to point out that this explanation doesn't make any sense, since we ice our water year round, even when it's freezing cold outside. But that sort of logical response just tends to make people angry, so now I just smile and keep my mouth shut.

At least one other sense tends to horn in on the whole issue of flavor, and that is vision. The appearance, or "presentation," of food is a critical element of its preparation in many cultures, and particularly at the finest of restaurants. In part 3, "Perception," we will visit a restaurant that goes to great lengths to ensure that its customers' appreciation of its cuisine is *not* influenced in any way by their vision.

## Supertasters

The term *supertaster* was coined by the psychologist Linda Bartoshuk in the early 1990s. A supertaster is defined, somewhat vaguely, as a person who experiences taste with far greater intensity than the average person. Researchers say up to one person in four of European descent may be a supertaster and that the trait is more prevalent in women than men and in Asians and Africans than in others. Various laboratory tests have been devised to determine whether someone is a supertaster. Such tests divide the population into three categories: supertasters (about one person in four), normal tasters (one person in two), and nontasters (one person in four).

One of the simpler of these rather qualitative tests involves a chemical called PROP. If a drop of PROP on your tongue does not cause you to sense a bitter taste, you are a nontaster. If you find the taste of PROP bitter but palatable, you are a normal taster. But if you find the taste of PROP so bitter as to be utterly revolting, you are a supertaster.

Supertasters often exhibit an abnormally large concentration of the small mushroom-shaped bumps that are found on the tips and sides of the tongue. Smaller than a pencil point, these growths typically contain three to five taste buds each. Other types of bumps, of different sizes and shapes and with different names, populate various regions of the tongue. Some of these contain taste buds as well, but most do not. They are there for other reasons, such as helping to break up small particles of food and mix them with saliva. The typical human tongue contains about five thousand taste buds, nearly half of which are located in deep grooves near the back of the tongue.

## *Signal Conditioning*

If it's not dissolved in water, you can't taste it. Your taste buds are no more stimulated by a solid chunk of sodium chloride (table salt) than they would be by a piece of gravel. Just as visible light is conditioned by the various optical mechanisms of the eye and sound waves by the outer and middle ear, so taste stimuli are conditioned by the machinery of the mouth.

When I was a kid, lots of things got me in trouble at the dinner table. Bad manners, pestering my younger brother—the usual childhood transgressions. Even in the act of eating, I managed to catch more than my share of grief. Turning up my nose at Mom's cooking was sure to raise my dad's ire, but so was eating too fast. Chewing your food carefully was important, Dad would say, because it is the first part of the digestive process. Chewing is also integral to the process of sensing flavor. Food is broken down when we chew it and at the same time mixed with saliva, which dissolves the salts, sugars, acids, and other molecules we can taste. Then the taste buds do the rest.

Most but not all human taste buds are located on the tongue. Smaller numbers are found on the roof of the mouth and at the back of the mouth, at the very top of the throat. Other creatures sport taste buds elsewhere. Many fish have external taste buds, allowing them to taste the water they are swimming through. A typical catfish might have 100,000 external taste buds, compared to the 5,000 on the average human tongue.

As with smell, the various taste receptors, on the tongue and elsewhere, are optimized for specific chemical stimuli. Salt receptors, unsurprisingly, respond to sodium chloride and to other salty chemicals such as potassium chloride. Receptors that respond to sweet tastes are stimulated by natural sugars and by artificial sweeteners.

As noted in the "Supertasters" section, there is a great deal of variation in "normal" individuals' sensitivities to various tastes. The physiological explanation is simple: some people have as many as one hundred times as many taste buds in the mushroom-shaped bumps at the front of the tongue as other people have. This fact corresponds well with the observed one-hundred-fold variation in threshold sensitivity to certain tastes.

## Taste Buds

In my junior high life sciences class, perhaps the same class where we did the apple-versus-potato taste tests, I learned that the tongue can detect four different kinds of tastes: sweet, bitter, sour, and salty. There is general agreement that a fifth taste category exists, which goes by both the exotic name of umami and the more common one of savory. Umami taste receptors respond to chemicals called glutamates. These occur naturally in foods such as tomatoes, meats, and cheeses. The best-known savory food additive is monosodium glutamate, or MSG. Some researchers report that certain taste receptors are selectively sensitive to still other categories, such as fats. There are important reasons why we can taste these things. Bitter tastes, for example, often result in revulsion, and not just for supertasters. And for good reason, because things that taste bitter are often noxious. But certain bitter-tasting foods, such as beer and coffee, are quite popular. Taste is more complex than it might at first appear.

Another lesson I learned in junior high, although it is false, is that the tongue is divided up into regions based on the four main taste categories. That misconception may have gotten its start decades ago through a poor English translation of a foreign textbook. Some areas of the tongue are more sensitive to one taste than another, the tip to sweet, for example, but all regions of the tongue are sensitive to all of the various tastes.

The human tongue is lined with bumps and grooves, as can be easily verified by looking in the mirror. Depending on the location, one of those grooves between the bumps might be lined with as many as 250 taste buds. Each taste bud contains around 100 taste receptor cells. Since there are about 5,000 taste buds in the average human mouth, that means there are about a half million (5,000 times 100) taste receptor cells. Compare that figure to the 3 million olfactory receptor cells in the nose, or the more than 100 million rods and cones in each retina.

The taste receptor cells are the cells that respond to molecules that taste sweet, salty, sour, bitter, or savory. Taste receptor cells are analogous to the rods and cones of the retina. They transform the stimulus into an electrical signal, and there are different receptor cells for each of the various tastes. What is the stimulus? That depends on which type of taste receptor cell we're talking about. The activation of a taste receptor cell by a tastant molecule is, at least in some cases, pretty well understood.

Consider table salt, sodium chloride, which activates the taste receptors that we perceive as "salty." Saliva on the tongue dissolves any salt not already in solution, and the sodium ions produced are able to penetrate the membranes at the end of the taste receptor cells. An increase in the concentration of positively charged sodium ions polarizes, or changes the voltage, of the receptor cell. That voltage change is the beginning of the chain reaction, as with vision, smell, and the other senses, that eventually results in communication to the brain and to the perception of a salty taste.

The stimulus reactions for other tastants are generally more complex than that for salt. Acids taste sour, but your tongue's acid taste receptors are not quite analogous to a pH meter, a device you might use to measure the acidity of, for example, swimming-pool water. A weak solution of acetic acid (the acid found in vinegar), tastes more sour than a weak hydrochloric acid solution of the same pH. Differences in the transduction mechanisms for various acids account for these results. Sugar and umami taste receptors utilize mechanisms that are more complicated still.

## Sugar and Everything Sweet

No other taste category has received as much attention from researchers, the food industries, and we eaters as the sweet taste. Vast sums of money continue to be invested in the search for low-calorie alternatives to sugar. Huge fortunes have already been made through sugar substitutes such as sucralose (Splenda) or saccharin (Sweet'N Low).

Cane sugar has a culinary history of more than five thousand years in Polynesia. After being introduced to the New World by Columbus and other Europeans, it flourished in the Caribbean climate. Europeans quickly developed a taste for this exotic product when it was exported back home. In 1700 the English were consuming 4 pounds of sugar per person per year. By 1800 the amount had increased to 18 pounds, and by 1900, to 90 pounds. In the United States, however, we take a back seat to no one in terms of our sweet tooth. In 2005 the average American consumed 140 pounds of natural sugars (cane sugar, corn syrup, and other natural sugars). That's 50 percent more than either the Germans or the French consume per person and nine times as much as the average in China. The average American sugar intake is roughly equivalent to 14 tablespoons of white granulated sugar per person per day.[3] Ugh.

[3]Since a tablespoon is a volumetric measure, it is difficult to accurately convert to a mass basis; the volume of a given mass of sugar depends on how fine the grains of sugar are. The 140 pounds of sugar per person per year converts to 174 grams per person per day. If a tablespoon of fine-grained sugar weighs 12.6 grams, that's about 14 tablespoons per day, or 2 tablespoons short of a full cup.

Based on those numbers, the average American consumes about 670 calories per day from sugar alone. Your recommended daily calorie intake varies depending on your gender, height, weight, age, and how active you are. There are lots of online calorie calculators you can consult. For many people, the recommended calorie intake per day is not far from 2,500. The calorie content in the average American sugar intake represents over one-fourth of that total. No wonder the obesity epidemic in the United States has such strong links to our intake of sugary foods.

How might we satisfy our sweet tooth without consuming all those calories? That's where artificial sweeteners come in, at least in theory. Those chemical concoctions have some pretty strange origins. Sucralose was developed in 1975 by a researcher trying to create a new insecticide. He happened to taste it by mistake and found that it was incredibly sweet. It was of no use as an insecticide, but, in products such as Splenda, it became the most popular artificial sweetener in America. Saccharin (1879) and aspartame (1965) were likewise products of serendipity. There is at least one good reason why each of these sugar substitutes was discovered by accident: it turns out that there is no general theory for determining whether a given molecule will taste sweet.

But that hasn't stopped people from trying to develop ever better substitutes for natural sugar. What makes one sugar substitute better than another? Lots of things. It has to be free from harmful side effects and thus worthy of acceptance by the Food and Drug Administration or its equivalent in other countries. It has to be really, really sweet. The new version of aspartame, called neotame, approved by the FDA in 2002, has been shown through psychophysical tests to be eight thousand times as sweet as sugar. But being sweet and benign isn't enough. It's also nice if you can cook with the stuff. Sugar is an astonishingly versatile ingredient in so many recipes. Unfortunately, some sugar substitutes, such as aspartame, break down at cooking temperatures, rendering them unsuitable for many pastries, candies, and other dishes. And finally, natural sugar has a complex taste that goes beyond simply being sweet, and it has very little aftertaste. Artificial sweeteners have to closely approximate these attributes, not an easy assignment.

It is also necessary for the taste of artificial sweeteners to endure for a long shelf life. Many artificial sweeteners break down over time so that their taste changes. If you've ever tasted a really old can of diet soda, you probably know what I mean. An engineer in the soda industry told me that the taste testers in his plant were hard pressed to tell the difference between the sugar-containing and artificial sweetener versions of the same beverage, provided the diet soda had just been produced. Within hours, however, the taste of the diet soda had changed enough that these trained tasted testers could distinguish it from sugar soda.

Neotame was the result of a twenty-year quest by the NutraSweet Corporation to replace aspartame. Teams of researchers on the two sides of the Atlantic adopted separate tactics for this challenge. An American team experimented with computerized molecular modeling. Starting with known sweetener molecules, they added a few atoms here and subtracted a few others there. The synthesized results of the computer models were fed first to mice, in order to weed out the deadly compounds, before employing human taste testers. Meanwhile, a French team adopted a lower-tech approach. Working without computer models, and tasting every single compound themselves, Claude Nofri and Jean-Marie Tinti eventually came up with neotame, which, as its name suggests, is a molecule whose chemical structure resembles aspartame. Neotame is sweeter than aspartame, and it has other advantages. For example, you can cook with it. It is on the market today, according to its Web site (neotame.com), in "over one thousand products worldwide."

The combination of science, intuition, and luck that characterized the search for neotame is gradually being replaced by a more straightforward scientific approach. This change is due is part to an ever-increasing understanding of how taste receptors work.

Since 1998, Charles Zuker has been searching for the genes that create the various taste receptors. He has identified genes that create receptors for sweet, sour, bitter, and umami. This knowledge has allowed Zuker and his team to breed what are known as "knockout" mice, which lack one or more of these various genes. Mice can be bred that lack the ability to taste bitter substances, for example, or sweet ones. By swapping genes around, it's even possible to produce mice who sense "sweet" when they taste something bitter, and vice versa.

This research has not only led to a better understanding of how taste works; it has also shined a light in the relative darkness where those who labored to develop things like artificial sweeteners worked. It has even resulted in the creation of a new industry—the taste potentiator industry. Taste potentiators are carefully engineered chemicals that, when added to foods in very small quantities, enhance a given taste characteristic. A few parts per million of a taste potentiator for the salty taste might allow a soup manufacturer to greatly reduce the sodium content in its soups without reducing the perceived saltiness of the soup. These compounds work in concert with whatever taste they are designed to enhance, such as saltiness. Just as a hearing aid turns up the volume of existing sound for the listener, so taste potentiators increase the taste sensation obtained from a given level of a certain ingredient.

One advantage of taste potentiators is that they are added in such small quantities that they need not be listed among the ingredients on the label of, say, a can of soup. They fall into the "other" category of "natural and artificial flavors" listed on some food labels. Not everyone is happy that taste potentiators can fly under the radar in this manner, but these substances do have to be accepted by the FDA before they can be used in foods.

Senomyx, a San Diego–based company, is a leader in this field. Senomyx has dispensed with the centuries-old role of the taste tester, replacing that person with a robotized, genetically engineered system that is breathtaking in its ability to generate truckloads of precise data about which specific types of taste receptor cells are stimulated by a given chemical. They use what have been called "artificial taste buds," which are tiny receptacles containing artificially grown human cells, all of a single taste receptor.

The receptacles also contain a fluorescent dye that illuminates whenever something reacts with the receptor cells. For example, if a drop of something sweet is added to a container of sweet receptor cells, the cells react, causing the fluorescent dye to light up. The light is then recorded by a photocell and the results fed to a computer. In this highly automated fashion, Senomyx tests millions of samples per year, a rate simply not practicable through old-fashioned taste testing.

Senomyx has a number of umami-taste potentiators on the market and has been working feverishly on sweet-taste potentiators—the Holy Grail of the industry. The company has marketed a sweet-taste potentiator that magnifies the effects of the artificial sweetener sucralose and has several that it claims are effective on natural sugar, although none are on the market as of this writing.

Before we leave sugar and other things sweet, note the following: Americans consumed about twenty-four pounds of artificial sweetener per person in 2005, a figure that had doubled in twenty-five years. During that same period, consumption of good, old-fashioned sugar also increased by 25 percent. The sweeter things become, it seems, the more we crave them.

# Chapter 7 • The Mechanical Senses

The mechanical senses are those whose stimuli involve mechanical energy, as opposed to electromagnetic or chemical energy. Mechanical energy can take various forms, but the mechanical senses are generally stimulated by the energy of motion—kinetic energy.

Vibrating air molecules contain kinetic energy, precisely because they are vibrating. The energy of motion in vibrating air molecules is transformed into electrical signals in the magnificent hardware of the ear. The motion or kinetic energy of the head, in particular, and the rest of the body is sensed by our balance and our proprioceptive systems. And the sense of touch generally involves the movement of one surface against another, for example, the fingertips of a blind person across a Braille manuscript.

## Hearing

Hearing just might be the Rodney Dangerfield of the senses. The late comedian "got no respect," and to some extent, that's the way it is with hearing, especially when compared to vision. I remember going to get my eyes examined many years ago. Proudly displayed on the wall in the waiting room was some sort of oath or pledge that the eye care professionals were making to their clients. They solemnly promised (I'm paraphrasing) "to do their utmost to protect the second-most-precious gift with which most people are born: their eyesight." The most precious gift, I was further informed, was life itself. It certainly wasn't hearing, which evidently ranked somewhere down the list, perhaps after neat handwriting.

If I'm a little cranky about the whole subject of hearing, please forgive me. It's Saturday afternoon, and I'm just back from my tennis game at my friend Bob's house. I lost, but that's got nothing to do with the foul mood I'm in. Bob lives in a lovely house with a beautiful tennis court. We started playing at 10:00 a.m., and at precisely 10:05, Bob's neighbor cranked up his riding lawn mower. The roar of the engine, even at a distance of fifty feet or more, was more than a little distracting.

We tennis players depend on our ears as well as our eyes. When your opponent hits the ball, your ears provide valuable clues about how well, or poorly, the ball was struck, and you react accordingly. If you don't believe me, try playing tennis while wearing earplugs. When those auditory clues are taken away, tennis becomes an even more difficult game. The great American star Jimmy Connors loved playing at the U.S. Open in Flushing Meadows, New York. There, the raucous crowds were nearly always behind "Jimbo" throughout his storied career, and their exhortations helped him pull through to victory on more than one occasion (he won the Open five times). But there was another thing Connors was said to love about the Open, and that was all the jets that flew in low over Arthur Ashe Stadium to and from nearby LaGuardia airport. The roar of the jets, like the din of Bob's neighbor's lawn mower, makes it very difficult to hear the ball strike the racquet. Connors was famous for his eagle eyes. His eyes were so good that he felt he didn't need the supplemental sensory input provided by his ears as much as his opponent did, and so he loved it when the jets roared overhead.

Bob's neighbor finished mowing the lawn about eleven o'clock, just as I was losing the first set of my match. Finally, silence on the court. Maybe now I could get into the swing of things. Not so fast. After you mow, you gotta blow, and out comes Bob's neighbor with his gas-powered leaf blower. Now I've really got problems. I'm losing the match and I still can't hear the ball, since the leaf blower is even louder than the lawn mower. What's worse, not only is the racket produced by the leaf blower "loud," as measured in decibels, but the blend of low-frequency noises (from the engine) and higher-frequency ones (from the air blower) is found by many people to be a particularly unpleasant combination. And I'm one of those people. So I'm losing my match, I still can't hear the ball, and the noise assaulting me is the lawn-care equivalent of fingernails on the chalkboard. Game, set, and match, Bob. Oh well, maybe next week Bob's neighbor will be on vacation.

I suppose I should lighten up and put my own problems in perspective. Helen Keller wrote, "I am just as deaf as I am blind. . . . Deafness is a much worse misfortune. For it means the loss of the most vital stimulus—the sound of the voice that brings language, sets thoughts astir, and keeps us in the intellectual company of man." These words were written by a woman who was either born deaf or became deaf at such a young age that she had no recollection whatsoever of the sensation of hearing.

The human ear is a marvel of electromechanical engineering. As a professor of engineering, I sometimes tell my students that one could organize an entire engineering curriculum around, for example, the automobile. Pretty much all the math and science that goes into the engineering curriculum is exemplified by one aspect or another of the modern car. But you could just as well organize an engineering curriculum around the human ear. Solid mechanics, fluid mechanics, vibration, hydraulics, controls, electrical circuits and electrochemistry, the properties of materials—all of these are well represented in the human ear.

Thermodynamics, the science of energy change, is importantly involved, too. The energy that is contained in vibrating air molecules—for this is what sound is—is minuscule. At the lower limit of hearing, the power produced by sound waves is about one-millionth of a watt per square meter. Sunlight, for comparison, can be a billion times more powerful at the earth's surface, producing up to 1,000 watts per square meter. The tiny amounts of mechanical energy contained in vibrating air molecules must be converted by the various mechanical, electromechanical, and electrical mechanisms in the ear into signals and sent to the brain. The depth of detail in those electrical signals is profound, such that we can instantly recognize the differences among a variety of musical instruments all playing the same note—a flute, an oboe, a clarinet, a trumpet, and a French horn, for example. And when a voice on the telephone says, "Hello," we instantly identify that disembodied, electronically reproduced voice as belonging to a certain one of scores, if not hundreds, of different individuals.

If you were an engineer told to produce a hearing instrument for a fancy new robot, and that instrument was required to perform on a par with the human ear, how would you begin? You might start out with the design specifications for the ear, to help you understand what the ear is required to do. First of all, the ear has to transmit sound over a wide range of frequencies. In the case of human beings, the audible range extends from about 20 Hz to about 20,000 Hz.

In other animals, the audibility range can be quite different. Cats and dogs hear much higher frequencies—dogs up to 45,000 Hz and cats over 60,000 Hz. Some dog whistles are thus inaudible to humans. Mice are able to hear up to 90,000 Hz but not below 1,000 Hz. Bats can hear sounds far above 100,000 Hz. These flying mammals emit high-frequency sound at up to two hundred pulses per second. Those sounds bounce off obstacles in the flying bat's air space and then echo back toward the bat. Bats apply this technique, known as echolocation, with such sensitivity that they can use it to detect the difference between a flying moth (food) and a falling leaf.

The human ear must also transmit sounds, whatever their frequencies may be, over a wide range of intensities. The more intense a sound wave, the louder we perceive it to be. Sound intensities, as discussed in part 1, are often represented using a decibel (dB) scale. By itself, the decibel scale does a poor job of conveying to a human being whether a particular noise is going to be pleasant or offensive, although it does indicate how loud we will perceive the noise to be. The human ear can detect sound intensities over a range of about 120 dB. That may not seem very impressive, but what it really means is that the energy level of the most intense sound we can hear (without severely damaging or destroying our ears) is about $10^{12}$ times (a million million times) more energetic than the least intense sound. The ear is both exquisitely sensitive to the faintest of sounds and robust enough to endure the shock of brutally loud ones.

The ear must also compensate for changes in atmospheric pressure and temperature. It has to function in both water and air. It must help us not only detect but also locate sounds—hence the need for two ears. And the ear has to keep on doing its job, with minimal performance degradation, for as long as the body in which it is installed remains alive.

## *The Hardware of the Ear*

As you the engineer continued your study of the ear in preparation for your design work on that robotic ear, you would progress from an understanding of what the ear has to do to an investigation into how it does it. Working your way from the outer to the middle to the inner ear, the technology keeps getting higher and higher, or the miracles more and more miraculous, depending on your point of view. There is ample room for both perspectives.

### THE OUTER EAR

Do older men have bigger ears than younger ones? Celebrities like H. Ross Perot and President Lyndon B. Johnson provide affirmative anecdotal evidence, and researchers have concluded that the external ear does indeed grow larger throughout our lives. More precisely, the external ear grows *longer* with age, for the width of the ear does not appear to change after its owner is full-grown. Ears lengthen by an average of 0.22 millimeter per year (about nine-thousandths of an inch) from physical maturity on. That may not seem like much, but it adds up. That growth rate would result in an ear that is more than a half inch longer in an eighty-year-old man than it was when he was full-grown at age twenty. The reason ears grow longer (both men's and women's ears) is more physical than physiological. It's gravity. The external ear, lacking bones to provide structural support, sags under its own weight and lengthens slightly with time.

Regardless of its size, the rather ridiculous appearance of the outer ear, that strangely shaped, almost comical assembly of cartilage and skin, belies the technical virtuosity that lies within. The outer ear consists of the auricle, or pinna, the part of the ear we see, and the ear canal (officially, the external auditory canal), which is about an inch long and a quarter-inch in diameter and curves inward to the eardrum. The eardrum seals off the inner end of the ear canal and, as its common name suggests,[1] acts like the skin of a drum to transmit the mechanical vibrations of the air molecules inward toward the middle ear and the inner ear, where things really get interesting.

## THE MIDDLE EAR

It is important for the air pressure on either side of the eardrum to be the same, and for that purpose the backside of the eardrum is vented to the atmosphere through the Eustachian tube. Discovered in 1563 by the Italian anatomist Bartolommeo Eustachio, this tube connects the middle ear, on the backside of the eardrum, to the windpipe and thus, most of the time, to the atmosphere. When atmospheric pressure shifts gradually, as when the weather changes or when you hike up a mountain, the Eustachian tube keeps up with things nicely and the pressure remains equal on both sides of the eardrum. But when the pressure on your eardrum changes more quickly, as when you're in a car that's going up or down a mountain or when you're ascending or descending in an airplane, the Eustachian tube can't always keep up. It wasn't designed for such high-tech modes of transportation. If the tube is partially or completely blocked, as it can be when you have a cold, and you have to travel by plane—well, most of us know the result. A splitting headache is accompanied by a distinct loss in hearing acuity. When this happens to me, it seems as if I'm underwater inside an aquarium, looking out and trying to understand what the people on the outside are saying. If only I could read their lips.

Modern airliners have pressurized cabins, but the pressure inside the cabin does not remain at atmospheric levels as the plane ascends to a

[1]The technical name of the eardrum is the tympanic membrane.

141

cruising altitude of 35,000 feet or more. Instead, in the ten or fifteen minutes it takes to reach cruising altitude, the pressure inside the plane reduces from atmospheric (sea level) pressure to the much lower pressure of an elevation of about 8,000 feet. Imagine if you could be transported—beamed à la Star Trek—from Miami at sea level to Vail, Colorado, at 8,000 feet, in fifteen minutes. This is what your ears are up against when you travel by plane, and this is what causes those splitting headaches when you travel with a head cold.

Aside from preventing headaches, the purpose of the Eustachian tube's role in equalizing pressures on the two sides of the eardrum is to allow the eardrum to function most sensitively and efficiently. If the pressure is not balanced, the eardrum's ability to transmit vibrations to the middle ear is diminished, and all of a sudden you're back inside that aquarium again, struggling to hear what's going on.

So, at the boundary between the outer ear and the middle ear, we have the eardrum. With pressure equalized on either side, it vibrates like a little drum skin as incoming sound waves bounce off its surface. But that's where the drum analogy breaks down, because firmly attached to the backside of the eardrum is—a bone. I'd like to have been in the design meeting when someone suggested that one. But in fact it is a great idea. The bone attached to the backside of the eardrum is often referred to as the hammer, although it is officially called the malleus. The three bones in the middle ear contain the answer to the trivia question "What is the smallest bone in the body?" The answer is the stapes, or stirrup bone, the smallest of the three. The largest, the hammer, is about 10 millimeters long, less than a half inch. The hammer is connected, at the other end from the eardrum, to the somewhat smaller anvil (incus), which is connected to the much smaller stirrup (stapes). The stirrup is only a few millimeters long, about half the size of a grain of rice. These three bones are known as the ossicles.

The ossicles are linked together to connect the eardrum, at one end, to the cochlea, at the other. Their job is to transmit into the inner ear all the frequencies and intensities of all the vibrations that impinge upon the eardrum, after amplifying them. The ossicles as a mechanical amplifier perform a crucial task. The amount of energy contained in sound waves is minuscule, as are the corresponding motions they cause in the eardrum. The higher the frequency, the less the eardrum moves. For sounds of very high frequency, which are still audible, especially to young people, the eardrum might move only about 0.05 nanometer, or just a fraction of the diameter of a hydrogen atom. Yet these tiny movements must be transmitted into the inner ear. It is only there that they can be transformed into electrical signals and sent on to the brain. In the strictest sense, "hearing" occurs only in the inner ear. The outer ear and the middle ear merely provide signal conditioning.

Airborne vibrations must thus be conveyed, through the eardrum and the ossicles, to the liquid-filled cochlea, where they are transformed into liquid-borne vibrations. This is no mean feat. The tiny forces in the vibrating eardrum must be multiplied (amplified) many times in order to reproduce those same vibrations in the cochlear fluid. The middle ear, with its three-bone linkage, makes this possible. Without it, hearing, at least for creatures that live on land, would not be possible.

The ossicles, the bones attached to the eardrum, have a little bit the appearance of a solution that was implemented to solve a problem that arose after the original system was designed. Engineers sometimes refer to things like this as a patch or a kluge. The original system, in this case, was a water-based hearing mechanism. The unanticipated problem arose when sea creatures migrated onto land.

Hearing evolved first in sea creatures. Hearing in water requires less complicated machinery than hearing in air does. This is because water-borne sound waves are much more easily transmitted into the liquid-filled apparatus of the inner ear. Air is a very light, highly compressible fluid, whereas water, more than eight hundred times as dense, is essentially incompressible. Hearing in water is simpler. Sound waves moving through water could be directly transmitted into the cochlear fluid, with no need for the amplification provided by the middle ear. When sea creatures evolved into land creatures, the middle ear followed.

The evolution of the mammalian middle ear, and in particular of the ossicles, is considered a classic example of the theory of evolution. Numerous examples of the transitional stages involved are preserved in the fossil record. The evolution of the ossicles into their role as a mechanical linkage between the eardrum and the middle ear is an example of exaptation, which is what occurs when anatomical features that developed for one purpose evolve into another role. Feathers, another example, originally served to help regulate the body temperatures of flightless creatures; later they evolved into their modern aerodynamic role in birds.

So it is with the ossicles of the middle ear. Clues to the process of change remain in the middle ears of some reptiles, where the eardrum is connected to the inner ear via a single bone, the stapes. The upper and lower jawbones of these creatures contain one extra bone each. These are the bones that, in mammals, have evolved into the other two bones in the middle ear.

Without a middle ear, airborne sound waves would impinge directly on the liquid-filled middle ear. A compressible substance, air, would be attempting to transmit energy into an incompressible one, the cochlear fluid. Engineers call this an "impedance mismatch." The result would be that very little vibration would get transmitted into the inner ear. Most of the vibration that impinged on the inner ear would be deflected away—it would simply bounce off the inner ear. The impedance mismatch between air and water is so great that the percentage of the acoustic energy in vibrating air that is transmitted into vibrations in water is very close to zero.

This is a show-stopping problem. Without a way around it, hearing in air is pretty much impossible. When I was a kid and went swimming with my friends, one of the goofy things we liked to do was to go underwater and then yell at each other. It was pretty easy to understand each other down there under water, in spite of the auditory interference supplied by all those hilarious air bubbles that escaped our lips along with whatever we were trying to say. But when someone else yelled at us from dry land when we were under water, we heard nothing—absolutely nothing. Sound waves propagating through air are no match for liquid.

This is where the middle ear and the ossicles come in. A mechanical engineer's dream, the ossicles serve to amplify the acoustic energy that arrives at the eardrum in two ways. First, there is a lever effect. "Give me a place to stand on, and I can move the earth," Archimedes is supposed to have said in describing the lever principle. With infinitely more delicacy, the ossicles multiply the force of the air pushing on the eardrum. Just how much mechanical advantage do they offer? For a simple lever, the kind Archimedes had in mind, the mechanical advantage is the ratio of the length of the lever on one side of the fulcrum to the length on the other. Let's say you want to move a 200-pound rock in your garden, and you have at your disposal a 6-foot-long steel bar and a cinder block. Place the cinder block 1 foot from the rock, lay the steel bar on top of the cinder block, wedge one end of the bar under the rock, and then push down on the other end. With the five-to-one mechanical advantage thus obtained, you can lift the rock by applying a 40-pound downward force on the other end of the steel bar, since 5 feet times 40 pounds equals 1 foot times 200 pounds.

The mechanical advantage afforded by the ossicles is much smaller. Calculating it is more complicated, since there are three bones in the linkage rather than just a single lever. The input force on the lever system formed by the ossicles is caused by vibrating air striking the eardrum, which is attached to the hammer. The output force, having been multiplied by a lever-factor of about 1.3 by the three-bone linkage of the ossicles, is applied to the oval window, which serves as the input to the cochlea. But the force per unit of area on the oval window is not just 1.3 times as great as that on the eardrum. That force per unit area, or pressure, is 22 times as great at the oval window as it was back at the eardrum. An entirely different force-multiplying effect is at work here, in addition to the lever effect.

The oval window is a tiny opening in the temporal bone that leads to the inner ear. With the important exception of the sound that enters the inner ear through conduction by the bones of the skull, virtually all auditory stimuli, everything we hear, passes through the oval window. The oval window is completely filled by the footprint of the stirrup bone, about half the size of a grain of rice.

That the oval window is so small is crucial to its success. The eardrum has a surface area about seventeen times as large as that of the oval window. All of the force spread out over the giant (in comparison) eardrum gets focused on the tiny oval window. Think of how a thumbtack works. The circular head of a thumbtack might be about 3/8 of an inch in diameter. You press on the head of the tack and apply a force of, say, 10 pounds. The pressure on the head of the tack is thus about 75 pounds per square inch (10 pounds divided by the area of the head of the tack). But on the sharp point of the tack, the pressure is multiplied 500 times, since the head of the tack is 500 times as large as the point. The pressure on the point of the tack becomes about 40,000 pounds per square inch. No wonder you can press the tack into a solid piece of wood so easily.

The pressure acting on the oval window is 17 times as great as that on the eardrum, due to the size reduction (the thumbtack or area effect). That 17-to-1 advantage is multiplied by 1.3 due to the lever action of the ossicles. Since 1.3 times 17 equals 22, there is a 22-fold increase in pressure from the eardrum to the oval window. This is perhaps the neatest parlor trick of the middle ear. But it is more than just a neat trick. Multiplying the pressure 22 times is enough to allow the feeble energy contained in airborne sound to be transmitted into the liquid-filled inner ear. As a result, those of us who hang out on land can hear.

Where are we at this point? The eardrum is vibrating, the forces are being transmitted through the ossicles, and the tiny stirrup bone is pumping pressure waves into the inner ear through the oval window. This is where things really get complicated. Those pressure waves pass through the liquid-filled cochlea, where they are transformed into electrical signals and then sent on, via the auditory nerves, to the brain. It is in the cochlea that the strictly mechanical actions of the outer ear and the middle ear are replaced by electrochemistry.

A transducer, as we have seen in discussing vision, smell, and taste, is a device that transforms one type of energy into another. Transducers can be biological or artificial. A car horn or stereo speaker transforms electrical energy into acoustical energy. A microphone works in the opposite direction, transforming sound into electricity. Is the cochlea just a microphone? Perhaps the cochlea fits the strict definition of a microphone, in that it converts liquid-borne acoustic pulses into electrical signals, but this flaccid description cannot do justice to one of the most magnificent organs in the human body.

### THE INNER EAR

The cochlea is often compared to a rolled-up tube, or a snail shell. The tube is rolled up for compactness, so that it packages better inside the skull, where space is at a premium. To understand how it works, first unroll the tube. But let's start with a different tube altogether, the one that Hungarian physicist Georg von Békésy (1899–1972) used in his work on the inner ear that won him the Nobel Prize in 1961. Actually, Békésy constructed quite a few different tubes in his efforts to understand the workings of the cochlea.

One of his most striking experiments, shown in figure 10, involved a water-filled tube about the same length and diameter as a human forearm. Békésy installed a thin, flexible membrane inside the tube, which ran the length of the tube, dividing it in half; the membrane was attached to a flexible ridge on the top of the rigid tube. He then sent sound pulses into one end of the tube. An observer, whose forearm was placed directly on the flexible ridge on the top of the tube, would then feel a gentle pressure point on his or her skin. But not all along the forearm. The sensation was felt in one small area, only an inch or so long. That pressure point would move from one end of the tube to the other as the intensity of the sound pulsed into the tube increased or as the frequency was changed. This is just one demonstration of what became Békésy's traveling-wave

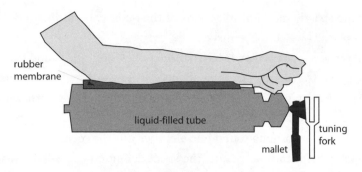

*Fig. 10.* Georg von Békésy's mechanical model, which helped explain the operation of the cochlea. (Adapted from Békésy, *Experiments in Hearing*)

theory of how the cochlea works. Essentially, the sound waves traveling through the fluid caused the flexible membrane to deflect, and the point of greatest deflection corresponded to where the observer felt the pressure point on her forearm.

The membrane running the length of the tube in Békésy's experiment mimics certain aspects of a similar flexible membrane in the cochlea. Attached to the membrane in the cochlea are thousands of tiny hairlike cells. When the membrane changes shape, when a traveling wave passes along its length, the hair cells move along with it. Each hair cell has one end embedded in the flexible membrane, which moves, and the other end is firmly anchored so that it can't move. Thus, waves traveling along the flexible membrane will pull or push on the hair cells. This is a key step in the transducer function of the inner ear.

An aspect of this process that was puzzling for a long time was how such a mechanism could account for the precision in our hearing, our ability to distinguish tiny differences in the frequency and intensity of sounds. Békésy's experiments went a long way toward explaining this as well. In his experiment, the skin on the forearm resting on top of the tube represents the hair cells, the sensory receptors in the ultrasensitive organ of Corti. In this way, Békésy not only demonstrated certain aspects of how hearing works, but he also showed that hearing and the sense of touch have

some things in common. Commonalities among the senses was another theme of Békésy's work.

The flexible membrane in the cochlea is more complex than the simple flexible membrane in Békésy's experiment. It varies in thickness and tautness as it extends from one end of the cochlea to the other. At the end nearest the oval window, it is at its thinnest and tightest. At the far end, it is thicker and looser. The thin, tight end transmits high-frequency sounds best, while the thick, loose end is optimized for low frequencies.

One way that Békésy's tube experiment differs from the cochlea is that the tube in the experiment is divided into two portions by the flexible membrane, whereas the cochlea is divided into three fluid-filled regions. There is a channel for pressure waves to enter the cochlea and another for those waves to depart. The third region contains the organ of Corti with its hair-cell receptors.

Pressure waves enter the cochlea through the oval window and pass along the length of the flexible membrane (see figure 11), where the process

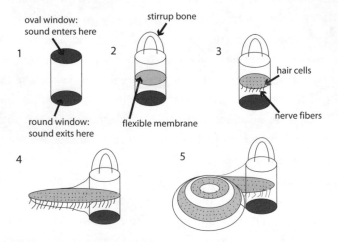

*Fig. 11.* How to build a cochlea: (1) Start with a tube that has a drum skin at top and bottom. (2) Add a stirrup bone to pump sound waves into the tube, and put a flexible membrane in the middle. (3) Add hair cells and nerve fibers to the flexible membrane. (4) Stretch out the flexible membrane sideways. (5) Coil it up to save space. (Redrawn and adapted from Nolte, *The Human Brain*)

of transduction begins. But then the energy in those pressure waves in the liquid has to go somewhere. At the far end of the membrane, farthest from the oval window, the pressure waves make a U-turn and return to the beginning end of the cochlea, through a separate liquid-filled passage. At the end of the return journey, the pressure waves are dissipated through a thin membrane known as the round window. By dissipating each successive pressure wave in this efficient manner, the cochlea ensures that successive waves are not acoustically contaminated by reflections from previous waves. This is analogous to the way things work in the eye: any light that gets past the rods and cones is absorbed at the back surface of the retina and is thus not allowed to bounce around inside the retina and confuse things.

The hair cells, anchored at one end to the flexible membrane, are contained inside the cochlea in the organ of Corti, an instrument of immense complexity packed into an almost impossibly small space. The organ of Corti is sometimes called the "seat of hearing." It is here that the mechanical vibrations of the flexible membrane are converted into electrical signals. The entire cochlea, rolled up, is about the size of the tip of the little finger, and it contains less than a teaspoonful of liquid. The cochlear duct, in turn, makes up only about one-sixth of the volume of the cochlea, yet it contains about seventy-five hundred separate moving parts. The exquisite delicacy of the organ of Corti requires the best protection the body can offer. The organ of Corti, inside the cochlear duct, is cushioned within the coils of the cochlea itself, which is buried deep within the temporal bone, the hardest bone in the body.

The organ of Corti is the transducer in the hearing process. Outward from the organ of Corti, all the energy is mechanical, as sound waves in the air are converted to vibrations in the eardrum skin, to the movements of the tiny ossicle bones, to pressure waves in liquid, and finally to the precise deflections of a delicate membrane. Inward from the organ of Corti, all the energy and all the processes are electrochemical, as the signals move deeper and deeper into the brain. The transformation

from mechanical to electrical energy is accomplished by those amazing hair cells.

## HAIR CELLS

When I was a kid, we listened to music on record players.[2] The way sound is produced by a vinyl record and a record player is reminiscent of the transduction process of the hair cells in the organ of Corti. (Very similar hair cells are also found in the organs of the vestibular system, responsible for our sense of balance, which is described a little later.) In a record player, a diamond stylus follows a groove that has been pressed into a vinyl record. The walls of the groove contain ridges. As the stylus passes through the groove in the revolving record, the ridges in the groove deflect the stylus back and forth. It is this mechanical motion of the stylus that is transduced into an electrical signal, sent on to an amplifier, and then retransduced into sound in a set of stereo speakers.

The pressure waves in the cochlea, analogous to the grooves in a record, cause the flexible membrane to deflect. The motion of the membrane results in the motion of the hair cells. Each hair cell, which might be 30 micrometers (a little over one-thousandth of an inch) in length by 5 micrometers in diameter, consists of a thick stock or trunk from which tiny fibers, called stereocilia, protrude on one end. An artist's paintbrush can be compared to a hair cell. The shaft or handle of the brush represents the shaft of the hair cell, while the business end of the brush, where the paint goes, represents the tiny fibers, or stereocilia. When the hair cell is deflected, it causes the bundle of tiny fibers on the end to deflect as well, sort of like a field of closely spaced wheat waving in the wind—but with crucial differences.

Each of the individual stereocilia is attached to the one next to it by an impossibly small and slender fiber called a tip link. Imagine that each

---

[2]My students tell me that vinyl records have made something of a comeback in recent years.

stalk of the waving wheat is attached to the next stalk by a tiny strand of spider web. In the ear, as the hair cell deflects in response to the motion of the flexible membrane, the tip links are stretched, opening tiny gates in the fiber. Opening these gates allows positively charged ions in the cochlear fluid, mostly potassium ions, to flow inside the hair cell and change the voltage of the cell.

So a pressure wave deflects the membrane, which in turn deflects the little fibers on the end of a hair cell. This motion stretches the tip links on those fibers. And that mechanically opens the channels that begin the electrochemical process that eventually sends an electrical signal to the brain.

The hair cells are the receptor cells in the hearing process, analogous to the rods and cones of the retina or the smell receptor cells in the nose. Whereas a retina contains 100 million or more receptor cells, there are only about 20,000 or so hair cells per cochlea. Of these, about 3,500 are inner hair cells, and the rest are outer. All of the hair cells, inner and outer, are contained in the organ of Corti, but they have different locations and they play different roles in the hearing process.

Those 3,500 inner hair cells per cochlea are responsible for what we call hearing. These are the sensory cells. The far more numerous outer hair cells function primarily as amplifiers. Inner hair cells are bristling with nerve connections to the brain. They pass along information about the frequencies and amplitudes of the sounds coming into the ear. By themselves, however, the inner hair cells are not nearly sensitive enough to react to most sounds in the normal audible range. The signals produced by the flexible membrane need to be amplified in order for the inner hair cells to detect them, and that's where the outer hair cells come in.

The outer hair cells don't really "hear" anything. That is, they don't pass along much auditory information to the brain. Instead, when an outer hair cell is stimulated, it responds instantaneously by *growing longer*. This length change amplifies the motion of the flexible membrane, which in turn increases the stimulus of the inner hair cells, allowing them to do their job. The amplification mechanism is crucial to proper hearing. Without it, for example if the outer hair cells of a cochlea have been destroyed by a drug, the affected ear will experience a staggering hearing loss of 60 to 80 dB.

Since the motion of the outer hair cells actually causes the flexible membrane to move, it would make sense that these motions would cause their own tiny pressure waves in the cochlear fluid and that a sensitive enough instrument, outside the eardrum, could detect them. This is in fact true, and it has become the basis for important hearing tests for newborn babies for whom hearing difficulties are suspected. A very sensitive microphone is inserted in the ear just outside the eardrum. A loud "click" is then played. If the sound makes it safely through the eardrum, the middle ear, and the cochlea, the amplification mechanism of the outer hair cells will cause a brief faint "echo" that is detectable by the external microphone. If no echo is recorded by the microphone, something is wrong with the middle or the inner ear.

## When Hearing Fails

Hearing loss comes with aging, excessive exposure to loud noise, and the use of certain drugs. It can also be caused by various diseases. A common mechanism for hearing loss is the death of the hair-cell sensory receptors in the organ of Corti. Hair cells in mammals cannot regenerate themselves. Once they're gone, they're gone. But the same is not true in all animals. The zebra fish, a tiny creature often found in freshwater aquariums, has the ability to regenerate the hair cells found on the outside of its body. It uses these hair cells to sense vibrations in the water—to hear underwater. Some songbirds possess a similar ability to regenerate hair cells in their ears. Researchers study these animals for clues that may aid in the treatment of hearing disorders in humans.

In the meantime, most folks who've lost appreciable numbers of hair cells, and thus some of their hearing, are forced to rely on hearing aids. The hearing aid is an amplification technique for the hearing impaired. There is no equivalent for vision or any of the other senses. Eyeglasses refocus light but do not amplify it. Situated in the outer ear, hearing aids amplify incoming sound waves, increasing the motions of the eardrum and the ossicles and boosting the pressure waves entering the cochlea. The hair cells that remain are able to take advantage of the increased stimulation and pass on a useful amount of information to the brain.

## The Cochlear Implant

Hearing aids are a practical solution to the problem of partial hearing loss. But what about a profound loss of hearing? What can be done for those who have lost all the hair cells in the organ of Corti? No amount of amplification can help, for the transduction mechanism, the means by which the mechanical energy in sound is transformed into electrical energy, is gone. In many cases, however, these individuals can be made to hear again. The solution is the cochlear implant, a device that bypasses all the magnificent hardware of the outer, the middle, and the inner ear and communicates directly with the brain.

A cochlear implant has five main components. Three of these are external to the body, while the other two must be surgically implanted in the patient's skull. The system takes in sound through a microphone (1) in a headpiece attached just above the patient's ear. This microphone does what a microphone always does—it converts sound into electronic signals. Those signals are passed along to a sound processor (2), which analyzes the signals from the microphone, using sophisticated software, and transforms them into a different set of electronic signals. The transformed signals are sent back to the headpiece, where a transmitter (3) passes them on into the skull, via radio frequency. There, surgically implanted internal electronics (4), sometimes referred to as the cochlear stimulator, receive the signal and transmit it as electronic pulses through a series of electrodes (5) that extend from the cochlear stimulator right into the cochlear duct and the organ of Corti. The electronic signals from those electrodes are picked up by the auditory nerves and then passed on to the brain.

The sound processor can be worn either on the person's belt or behind the ear like a conventional hearing aid. In either case, its role in the overall function of the cochlear implant cannot be overstated. The way it transforms the electronic signals from the microphone determines what the person will hear. The software contained therein is crucial to the perception process. Different software packages exist, and people who have received a cochlear implant often report great differences in what they can and cannot hear, depending on which software package is employed.

Cochlear implants have restored the hearing of many profoundly deaf persons. But what those individuals hear, with their cochlear implants, is vastly different from what they heard before they became deaf. Cochlear implants transform sound waves into electrical signals that the central nervous system then passes on to the brain. Cochlear implant recipients such as Michael Chorost, whom we met in the beginning of part 2, report that what they hear, what their brains perceive, when their cochlear implants are first activated, regardless of which software package the sound processor utilizes, is little more than an unintelligible cacophony. The story of how the brain learns to interpret those new and different signals is recounted in part 3, "Perception."

## Turn That Thing Down!

One of the main causes of hearing loss is, most of the time, utterly preventable. What do iPod listeners, rock musicians, and lawn-care workers have in common? Each potentially subjects herself to unhealthy, damaging levels of noise.

Unwanted noise, or noise pollution, combines aspects of stimulus, sensation, and perception. The word *noise* can refer to any kind of sound, but it is most often used for sounds that are "loud, unpleasant, unexpected, or undesired," according to the *American Heritage Dictionary*. Whether a sound is unwanted often depends on who is hearing it. Loud music in the middle of the night is perceived differently inside the apartment where it is being played than it is next door, where folks are trying to sleep. Noises can do things to us psychologically, as that example shows. But they can also damage us at the sensation level, in the hardware of hearing.

The effects of prolonged exposure to noise at various levels are becoming better understood all the time. In terms of damage to your ears, it matters how loud a noise is, but it is also very important how *long* those loud noises pound your eardrums. Unless the noise is exceptionally loud, it's not your eardrums that suffer the effects of long-term loud noise, so much as it is those amazing hair cells. It used to be thought that the degradation of the hair cells brought on by prolonged noise was essentially a mechanical effect. It was believed that, much as the ocean waves crashing ashore, day after day, year after year, eventually wear down a rocky cliff, hair cells would mechanically break down after so many hundreds of hours of running that leaf blower, riding that Harley, or cranking that iPod up to "eleven." It is now known that the mechanism by which prolonged loud noise degrades hair cells is biochemical, not mechanical. Prolonged loud noise causes the electrochemical reactions in the hair cells to fire more or less continuously. Eventually, this leads to the degradation and death of the hair cells. Michael Chorost says the hair cells are "committing suicide." I'm not sure how apt that description is, since the hair cells themselves didn't decide to expose themselves to whatever prolonged racket it was that caused their death.

If a noise is loud enough, a single exposure, no matter how brief, is enough to damage hearing. The threshold for single-exposure damage varies depending on whose data you consult. Anything above 140 dB is to be avoided resolutely. What sorts of things are that loud? One example would be an airplane's jet engine under takeoff thrust. Depending on the type of engine, noise blasts in excess of 140 dB are possible within a hundred feet or so of the engine. Workers who launch professional fireworks can be exposed to similar noise levels. And there are plenty of car stereos capable of emitting sound at a level of 140 dB.

It's easy enough to avoid superloud car stereos, but you can't always control things where you work. Lots of workplaces are noisy enough to damage one's hearing. As an engineer, I've worked in or visited many of them. The shriek of a high-pressure letdown valve in a petroleum refinery is particularly painful and unpleasant. I also remember touring a plant where a nail-making machine was used. You feed steel wire into it, and the machine chops the wire up into nails at very high speeds, accompanied by a truly infernal racket.

OSHA, the Occupational Safety and Health Administration, which has long recognized the dangers of noisy workplaces, publishes regulations that U.S. employers must follow. OSHA regulation 1910.95 (a) includes a table showing the maximum permissible noise levels, as measured on the dBA scale and averaged over an eight-hour workday. As long as the average noise level is below 90 dBA, workers can put in an eight-hour shift with no hearing protection. At 95 dBA, the maximum daily exposure is four hours. At 100 dBA, it drops to two hours, then to one hour at 105 dBA, thirty minutes at 110 dBA, and 15 minutes at 115 dBA. How loud is 115 dBA? If you're near the stage at a rock concert, there's a good chance the music reaches your ears at 115 dBA or better.

The OSHA limit for all-day exposure, 90 dBA, is plenty loud. Other organizations place the limit for continuous exposure at 85 dBA. Lots of common household items are that loud or louder. Virtually any handheld electric tool, such as a drill, a saw, or a grinder, exposes the user to noise in excess of 90 dBA, as do our old favorites the lawn mower and the leaf blower. Even kitchen implements such as blenders and coffee grinders are this loud. If you use any of these tools with any frequency, you should consider wearing hearing protection.

Millions of us go about our daily routine with tiny earbuds poked into our ears, pumping digitized music into our brains at what are often dangerously high noise levels. The percentage of teenagers suffering from hearing loss has greatly increased in recent years, and much of the blame has been heaped upon the loud music we pipe into our ears via earbuds.

By the time I graduated from college, my modest collection of vinyl LP records had grown to several hundred. Moving them from place to place was a real pain. Nowadays all of that music, and plenty more, is easily stored in the palm of my hand in a device smaller than a pack of matches. Among the greatest advances in digital music have been so-called compression technologies, such as MP3, that allow recorded music to be digitally encoded. The resulting electronic file is of a manageable size and thus amenable to computer downloading, storing on a portable player such as an iPod, and so on. What was once an all-consuming quest for fidelity has been replaced by a similar one for convenience. But there's more to MP3 than that. The music is compressed in such a way that it is efficiently stored electronically without significantly compromising sound reproduction quality.

These days, we're all used to the idea that you can take all of your music with you, no matter where you go. But taking all that music with you, and listening to it, is not without its perils. Portable music players such as iPods that utilize earbuds typically have maximum volume levels of 100 dBA or higher. OSHA limits exposure to 100 dBA work environments to no more than two hours a day without ear protection. It is thus not hard to imagine that many listeners are probably exposing themselves to harmful levels of noise.

Various regulations in Europe restrict the volume on portable music players. Devices sold in France are limited to 100 dBA maximum. European Union regulations require the players to default to 85 dBA each time they are turned on, although that volume level can be changed by the user.

The effects of noise pollution at levels below OSHA and other thresholds are not as well understood and are consequently more controversial. But noise doesn't have to be damaging to a person's hearing mechanism in order to be dangerous. We return to this subject in part 3.

## Touch

Although it is up against some serious competition, I hereby nominate the star-nosed mole as one of the craziest-looking critters on the planet. And its star nose would certainly merit honorable mention, or better, for the title of most unusual sensory organ in the animal kingdom—alongside the narwhal's tusk. Star-nosed moles, which live underneath some wetlands in eastern Canada and the northeastern United States, spend most of their time burrowing through mud searching for insects and worms. Like other moles, they have powerful, scary-looking front paws that serve as efficient

*Fig. 12.* The star-nosed mole, with eleven tentacles surrounding each nostril. The mole's tiny eyes, on either side of the star nose, are not visible in this photo. (Photograph by Mary Holland; used by permission)

earthmovers. The star nose protrudes between tiny, feeble eyes. Each nostril is surrounded by eleven pink tentacle-like growths, as shown in figure 12. The tentacles are only about one-eighth inch long, but these are diminutive moles, generally weighing less than two ounces. They can be up to eight inches long, but three of those inches are tail. On an animal that size, the star nose is indeed prominent.

One might guess that the tentacles, given their proximity to the nose, are somehow involved with taste or smell. Au contraire. As sense organs, the tentacles are more like fingers. They are for *feeling* the earth the mole is burrowing through, not smelling or tasting it. The tentacles do not grasp objects or perform other fingerlike functions. They are simply highly specialized tactile sense organs, vibrating through the earth at frequencies up to 10 Hz as the mole is searching for food. The eleven tentacles around each nostril have about 50,000 nerve endings devoted to the sense of touch. By contrast, the much larger human hand contains only about 17,000 touch-sensitive nerve fibers.

Dogs, with their 200 million or so smell receptors, have olfactory capabilities far greater than those of humans. Based on the nerve endings possessed by star noses, are we to conclude that star-nosed moles are similarly superior in tactile ability? Perhaps, but it hasn't been scientifically confirmed. To find out, someone needs to run experiments on the spatial-resolution capabilities of the star-nosed mole's tentacles.

Such experiments have been carried out on humans. They confirm what you probably already suspected—that the fingertips are the most touch-sensitive parts of our hands. Touch sensitivity can be quantified in terms of how much distance between two stimulation points is necessary to recognize them as separate points. It's a pretty easy experiment to reproduce, at least to get a rough idea of what's being measured. Grasp two toothpicks with your thumb and forefinger so that you can control the distance between the ends of the toothpicks. With the ends about an inch apart, poke them gently into the center of the palm of your other hand. You should feel two separate pressure points; that is, it should be obvious that there are *two* toothpicks. It helps if you look away while doing this, or get a friend to wield the toothpicks, so that you can keep your eyes out of the experiment. Now move the ends of the toothpicks quite close together, say about a millimeter apart, and repeat the experiment. The palm of your hand isn't nearly sensitive enough to distinguish two stimuli so close together, and you will sense this as a single point. If you gradually increase the distance between the ends of the toothpicks as you repeatedly poke yourself, you will eventually arrive at a separation distance just large enough to sense that there are two toothpicks, not one.

Researchers have shown that the minimum separation distance on the palm of the hand to enable the average person to sense such inputs as separate is more than 8 millimeters (just over 0.3 inch). The fingertips are much more sensitive. There, the minimum separation distance is less than 2 millimeters. Hold the toothpicks 3 or 4 millimeters apart, and poke yourself first in the palm of the hand, then on a fingertip. It should feel like a single stimulus on the palm but like two on the fingertip.

## Touch Receptors

The results of those touch sensitivity experiments match quite well with the distribution of touch-sensitive receptors in the skin. At the fingertip, there are more than two hundred touch sensory receptor units per square centimeter, about four times as many as in the center of the palm of the hand.

Two hundred receptors per square centimeter means that on the tip of your finger an area the size of the head of a pin contains about six of these touch receptors. Because the senses we associate with our skin, such as touch, temperature, and pain, appear to us to be continuous over our skin, it is easy to imagine that the receptors responsible for these senses are somehow homogeneous. The toothpick experiment indicates this isn't true; rather than uniformly covering the skin, the receptors are distributed like dots. The farther apart the dots are, the less sensitive we are in the areas between the dots. If a slender probe is chilled and then touched to the skin, it will feel cold—if it is close enough to a temperature receptor on the skin. If, by chance, the cold probe is far enough away from the nearest temperature receptor, it is likely that the only sensation felt will be touch.

## Touch, and Other Things, in Robots

When we think of the robots of the future, we often think about brains. Robots will be *so smart*, someday. The robot HAL in Stanley Kubrick's *2001: A Space Odyssey* comes to mind. Advances in robotic intelligence are impressive, but that's not the only thing robots are getting really good at. The sensory capabilities of some robots, even today, are breathtaking.

Three-fingered robotic hands with incredible dexterity, speed, and tactile sensitivity have been developed, and the parlor tricks they can perform are astounding. One such three-fingered robotic hand can pick up a cell phone, toss it in the air, and then gently catch it as it falls. Required are a combination of robotic vision, to sense the position of the falling object; touch; and proprioception, to successfully catch the phone without damaging it. The same robotic hand is capable of tying knots in thin cords. It can also grasp a pair of tweezers, use them to pick up a grain of rice, and then place the rice in a small cup.

When I showed my wife a video of the robotic hand I just described, she was duly impressed. It really is pretty cool. But her first question was,

"What could you do with a robot hand like that, besides all those tricks?" One answer is robotic surgery.

Robotic surgery is not entirely new. Its first use on a human being dates to 1965. Advantages often touted for robotic surgery include the ability to perform remote surgery, so that the doctor controlling the robot could be on the other side of the planet, and the minimally invasive nature, in general, of robotic surgery compared to conventional surgery. "Minimally invasive" generally refers to laparoscopic surgical techniques, robotic and otherwise, that can be accomplished with very small incisions.

The doctor is always in charge during robotic surgery—autonomous robotic surgery remains the stuff of science fiction. What the robot offers is an end effector, the part of the machine that wields the surgical instruments and also positions the internal video camera, that has a greater range of motion than the human wrist. The robot can also move the surgical instruments with more precision and less tremor than a surgeon can. However, a surgical robot is likely to cost on the order of a million dollars, with hundreds of thousands of dollars in maintenance costs per year. Studies comparing robotic surgery to conventional laparoscopic surgery don't always give a clear victory to the robots.

In terms of the senses, robotic surgery is purely visual. Robotic surgery is just a highly sophisticated video game, albeit one with life-and-death consequences. Time will tell whether robots with other sensory capabilities, such as touch, will be employed to make the game even more sophisticated.

## The Pain Response

The skin and the regions just below it are teeming with sensory receptors. We've just visited the sense of touch, which is related to our abilities to sense pressure, vibration, and other stimuli, including those that provoke an itching sensation. The sensations received in the skin, in the flesh just below it, and in the deep tissues of the body are sometimes categorized separately, to distinguish them from the specialized sensory organs such as the eyes and the ears. Such sensations thus include, for example, the receptors that reveal to what extent a muscle is flexed, or whether your stomach is full.

The concept of sensing pain, or nociception, was introduced in part 1, where we noted that different receptors are responsible for the pain response than for other sensations. Recall the example of dipping a finger into warm versus hot water. Both provoke a temperature sensation, but only the hot water provokes pain.

Pain receptors are different from other receptors in at least one other way. Stimuli that would not normally be perceived as painful can be decidedly so, if the affected area has been injured in some way. Typing on a keyboard, for example, is generally a pain-free experience, but not if you have a small burn or cut on the end of a finger. Each time the offending digit touches a key, a sharp pain is sure to follow. Pain receptors must have some sort of memory for areas of the body that have been damaged. In fact, pain receptors can both send information to the brain and receive information back from the brain.

Information in the form of electrical signals flows to and from many parts of the body. Muscles are a good example. When you want to pick something up, your brain sends information to the muscles of the arm and the hand, but the act of picking something up also requires information to be sent from those muscles back to the brain. The brain needs continuous feedback to accomplish tasks like this, as we shall see when we discuss proprioception, or body awareness.

Most sensory receptors only send information to the central nervous system; they don't receive it. Pain receptors are different. When the body is injured, a variety of processes are set in motion to begin the healing process, and pain receptors play a role. The same receptors that sent the brain the pain signal in the first place now aid in healing the affected tissue. The example of a small burn on the skin has been well studied. The cells damaged by the burn release chemicals that hypersensitize the endings of the pain receptors in the region. This is why the area around the burn is so painful. Pain receptors are also involved in the processes that make the burned area turn red and swell up.

Most sensory receptors, such as the rods and cones of the retina or the hair cells of the inner ear, are highly specialized anatomical constructs. Pain and temperature receptors, however, are essentially just free nerve endings. They respond to different stimuli, but it isn't clear what it is about their structures that make them respond differently.

Pain receptors can respond to painful levels of hot and cold; to painful mechanical stimuli, such as a cut, a prick, or a pinch; and to many of the various chemicals noted above that are released when tissue is damaged. Some pain receptors respond to all of these stimuli, others to only one or two kinds.

Yet another factor differentiates types of pain receptors. When the nurse pricks your finger for a blood sample, the first thing you feel is a sharp pain. What follows soon afterward is a dull ache. These two types of pain have been traced to different types of pain receptors, whose actions have been verified by sophisticated experiments. Special anesthetics can block either the dull-pain or the sharp-pain receptors, so that a person whose finger is pricked feels only the type of pain that has not been blocked.

Most of us have probably wished at some point in our lives that we could not sense pain. I remember a particularly exquisite dental abscess that made me feel that way. Pain exists for very good reasons, though. Individuals who are incapable of feeling pain are rare, but they do exist. Not only do such persons, lacking the body's warning system, tend to suffer many more injuries than the rest of us, but those injuries tend to heal poorly or not at all. Pain receptors, after all, both send pain signals to the brain and aid in the healing process.

## Temperature Receptors

Temperature receptors, or thermoreceptors, are, like pain receptors, free nerve endings without any obvious specialized structure. Temperature receptors respond to changes in temperature by opening or closing channels in the nerve that allow positively charged ions to flow, much like the various other receptors we've discussed. Studies reveal the existence in humans of at least six different types of temperature receptors, differing only in the temperature ranges that activate them. These receptors are most sensitive to the temperatures 50, 59, 95, 104, 113, and 140°F (10, 15, 35, 40, 45, and 60°C). The first two are thus "cold" receptors, the next two "warm" receptors, and the last two "hot" receptors. The pain threshold of the average person is around 113°F. The connection between temperature sensing and the flavor of food was discussed earlier. The same channels that are opened by changes in temperature are also opened by certain foods such as chili peppers and menthol, which we refer to as "hot" and "cool." So it's no coincidence that the words we use to describe the sensations these foods create are the same ones we use for temperature effects; in each case, the same receptors are involved.

In part 3, "Perception," we will discuss the blind spot in vision and other examples of what is known as perceptual completion. These are ways in which the senses can be fooled, and they include our ability to sense temperature. Take three quarters, and put two of them in the freezer for about five minutes. Then line up the three quarters on a table, with the room-temperature quarter in the middle. Touch all three quarters simultaneously, with the tips of your index, middle, and ring fingers. The perception that *all three* quarters are freezing cold is striking. If you touch *only* the center quarter, it will not feel cold. Touch all three at the same time, however, and they will all feel cold.

# Balance

I love to go to the movies. I don't love everything about the experience, though, and at least one cinematic trend these days sometimes has me wishing I'd stayed home with a good book. The 2009 film *The Hurt Locker*, which won the Academy Award for Best Picture, is a riveting tale of a team of American soldiers in Iraq whose job it is to find and disarm improvised explosive devices (those infamous IEDs). I could hardly watch it. Not because the film is so intense, which it is, but because of the widespread jerky hand-held camera work. I suspect the filmmakers used this technique because they believed it gave the moviegoer a sense of really being there. In *The Hurt Locker*, it does exactly that. You *are* right there, creeping down a Baghdad alley, not knowing which parked car or pile of trash contains a bomb, or which rooftop conceals a sniper. When the camera work is steady, perhaps the viewer feels more detached. But when the camera jerks around, you feel as if you're down there on that dusty street with the soldiers. Except for me. I feel as if I should be heading for the men's room, to throw up. More than a few minutes of jerky camera work, and I am overcome with feelings of nausea akin to seasickness. The Motion Picture Association of America affixes ratings to movies (G, PG, R, and so on) as a guide to viewers. Might I suggest the addition of W, for Wobbly, to aid viewers like me?

Seasickness, the kind you get on a boat, is caused by a combination of visual stimuli (the heaving deck, the wobbling mast) and mechanical stimuli, as the body is jerked about by the action of the waves on the boat. Both of these, the visual stimuli and the mechanical acceleration of the body, are tied in to our sense of balance. Some folks can get seasick from jerky visual stimuli alone, and the cinematic evidence suggests I'm one of them. It's not much fun. In his book *The Human Body*, Isaac Asimov recounts the tale of a seasick passenger during an ocean voyage. A steward attempts to reassure him by telling him that no one ever died from seasickness. "It's only the hope of dying that's keeping me alive," responds the poor passenger. Seasickness is an unfortunate manifestation of our sense of balance, an otherwise indispensable sense, whose specialized hardware rivals that of vision and hearing in its sophistication.

## How Do We Sense Position and Acceleration?

If a vote were taken, I suspect the vestibular system would win the award for "least known and most underappreciated sense organ." Most of us are aware that we have something called a "sense of balance." We may realize that this sense has something to do with the inner ear, but beyond that, many people haven't a clue, at least based on my informal polling.

One problem is that, unique among the sense organs, the vestibular system, which senses linear and rotational accelerations of the head, is entirely internal. The eyes, the ears, the nose, the skin, and the tongue can all be seen. Could this be a case of "out of sight, out of mind?" When we catalog the senses for a young child, each of the other sense organs can thus be pointed out. But I've never seen anyone complete such a tour by noting, "Oh, and one more thing, Sally. In addition to all those wonderful organs, you've got a set of semicircular canals and otoliths in each inner ear. They keep you from falling down when you stand up and walk around."

But those organs are there, and that is what they do, among other things. Life as we know it would be impossible without them. The sense organs that contribute to our sense of balance and orientation are collectively known as the vestibular system. This system includes ten separate accelerometers, five in each inner ear. Six of the ten sense the rotation of the head, while the other four sense the linear motion and position of the head.

## Sensing Rotation of the Head

The six rotation sensors take the form of hollow, fluid-filled rings called semicircular canals. They look somewhat like a baby's teething ring. Each ear has three semicircular canals, as shown in figure 13. Each of the three canals is roughly perpendicular to the other two.

One canal lies in a plane tilted down about 30 degrees from the horizontal. Looking at a human head in profile, it lies in a plane sloping down

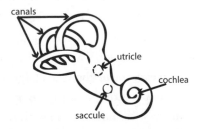

Fig. 13. The vestibular system, showing the three semicircular canals and the utricle and the saccule, in relation to the cochlea in the inner ear.

from the level of the eyes toward the back of the neck, the lower end of it. The other two canals are in vertical planes. One of the two in my left inner ear would look like an "O" if you were standing in front and to my left, while the other would look like an "O" if you were standing behind me and to my left.

As you may have guessed, having three such mutually perpendicular canals allows the body to sense rotations about each of three axes. The canal closest to the horizontal plane detects rotations of the head about the vertical axis, the type of motion you make when you shake your head "no." No matter which way your head is rotating, at least one of the three canals in each ear is oriented optimally to sense it.

Here's how the semicircular canals work. Imagine that something catches your eye on your left. You quickly rotate your head in that direction. The semicircular canal, being rigid, rotates right along with the rest of the head. To make things simpler, pretend that you have only one such canal. The canal rotates, but the fluid inside it does *not* rotate with it—at least not right away. It's a little like what happens when you make a sudden move while holding a beverage in your hand. The solid beverage container follows your hand's motion just fine, but the beverage inside the container, being a liquid, can't do the same. It sloshes about, and you end up with coffee on your white shirt. The sloshing motion of the fluid inside the rigid canals is the key to their ability to detect movements of the head.

The fluid motion inside the semicircular canals also reminds me of a trick cooks use to tell whether an egg has been hard-boiled. Place the egg on its side on a flat surface, and spin it like a top. If the egg spins freely and rapidly, it's hard-boiled. But if the egg makes just a few half-hearted rotations and then slows to a stop, it's raw. Roughly the same principle is at work inside the raw egg and inside your semicircular canal or your coffee cup. When you spin an egg, you're giving it a large angular acceleration: it quickly changes from not rotating at all to rotating fast. If the egg is raw, the gooey liquid inside it cannot keep up with the motion you've just given the rigid shell. When the shell begins rotating in one direction, the liquid inside it sloshes back in the other direction, where it bounces off the wall of the shell. All this extra motion of the liquid sloshing around inside the shell absorbs most of the energy you imparted to the egg by spinning it, and as a result the raw egg spins only a few times before it poops out and stops. The hard-boiled egg, by contrast, contains no liquid; when you spin it, the whole thing, the shell plus its solid contents, follows the spinning motion you supply, and the egg spins freely.

So the raw egg, when spun, acts like the liquid-filled semicircular canal. When you turn your head, thus rotating the canal, the liquid can't follow the motion right away. If the ring-shaped canal is rotating counterclockwise, the liquid inside it sloshes back clockwise. And now we finally get to the point of all this. Inside the canal, the liquid is free to travel all the way around, 360 degrees, except at one point. At that point, there is a wide spot in the canal called the ampulla, for its blisterlike appearance. Stretched across the ampulla is a flexible membrane, which prevents the fluid in the canal from passing. When you turn your head, the canal turns with it, but the fluid inside sloshes in the other direction. Almost instantly, the fluid slams into the flexible membrane, deforming it. It is that deformation that you sense, because the membrane contains hair cells that operate in a manner similar to that of the hair cells in the cochlea. This "backwash" of the fluid in the canal tells the brain that the head has begun to turn. Let's say you rotate your head only 30 degrees and then abruptly stop it. The fluid in the canal will help you sense this as well. The backwashed fluid bounces off the membrane and starts moving in the same direction as the canal. But when the head and the canal stop moving, the fluid keeps going, eventually slamming into the membrane from the other direction. This alerts the brain that the head has stopped moving.

These pulses of fluid, always lagging just behind the motion of the canal, send signals to the brain regarding the start-and-stop motion of the head. You can see that what the semicircular canals excel at is detecting *changes* in motion—starting and stopping. After all, our heads do an awful lot of this sort of start-and-stop movement. By contrast, the semicircular canals are poor at sensing *continuous* rotation, such as what you might experience on the tilt-a-whirl at an amusement park. This is one reason why such rides are so disorienting, not to mention nausea-inducing.

This description might give you the impression that the vestibular system reacts rather slowly. Far from it. Our balance reflexes are among the fastest we possess. If you shake your head while reading these words, for example, you will notice that your eyes have no trouble maintaining their focus on the words. This is known as the vestibulo-ocular reflex, a collaboration between the vestibular balance system and the muscles that position the eyes. To enable this reflex action to occur as quickly as possible, the nervous connections involved are simple and direct. Thus, eye movements lag behind those of the head by only about 10 milliseconds. Shake your head slowly, as if you were saying, "no," while reading these words. The words on the page should remain just as stationary as they are when you are reading normally. Now, shake your head as fast as you can. If you're like me, you can still read the words when you do that, but there will be a slight "bounce" to them. Shaking my head that rapidly moves it a little faster than my vestibulo-ocular reflex, so my eyes lag just a tick behind the motion of my head.

Just as we have two eyes and two ears, we have two vestibular systems, one in each inner ear. Each contains a set of three semicircular canals. As with the eyes and the ears, more is involved than just redundancy. Each left-right pair of canals operates in what is known as a push-pull arrangement. That is, when one canal, say on the left, is stimulated by a head movement, its counterpart on the right is inhibited, and vice versa. Doctors can take advantage of this feature when examining a patient for problems related to balance. In the "rapid head impulse test," a doctor asks the patient to focus on the doctor's nose. The doctor then carefully but rapidly rotates the patient's head either left or right about twenty degrees. If the patient is able to maintain focus on the doctor's nose, no problems are indicated. But if, for example, the patient's eyes react normally when the head is rotated to the left, but not when it is rotated to the right, a problem in the right vestibular system, perhaps due to a disease or an accident, is indicated.

## The Segway

Dean Kamen, whose company DEKA invented and now markets the Segway, is an engineer's engineer. His technical brilliance and creativity are legendary, and he combines them with a savvy business sense. His success has allowed him to give back to the community, which he does with a sort of inspired generosity that is rare indeed. While it has always seemed to me that his Segway is the answer to a question that no one was asking, there's no doubt that it is an amazingly sophisticated, utterly cool machine. Kamen has said he was inspired by the human body's balance control system when he invented the Segway. Instead of walking on two feet, the Segway rolls on two side-by-side wheels, but the way it controls itself to keep from falling over has some things in common with the way we keep ourselves upright when we walk. The Segway has both a sense of balance and a sense of proprioception.

The Segway is powered by two independent electric motors, one for each wheel. The motors can turn each wheel forward or backward, so they also act as brakes. The motors are controlled by a computer connected to a set of sensors that mimic the body's vestibular system. These sensors, essentially miniature gyroscopes and accelerometers, measure the angular position and acceleration of the frame of the Segway.

When the driver of a Segway leans forward, the gyroscopic sensors detect a rotation of the frame in that direction. To keep the machine (and driver) from falling over, the computers signal the motors to power the wheels forward, and the Segway begins to move. For forward and backward motion, that is how the driver controls the machine—by leaning in one direction or the other. The wheels turn just fast enough to keep the machine upright and balanced. There is continuous feedback, hundreds of times a second, between the sensors, the computer, and the motors.

Walking works in a similar way. When we walk, we lead with our head. We lean forward, and our legs automatically move, one after the other, to maintain an upright posture as we walk down the street or across the room. Our vestibular system senses the position and acceleration of our head and sends corresponding signals to the brain, which directs the legs to get moving. The more we lean our head, the faster the legs have to move. If we lean far enough, we have to transition from walking to running to keep up. The Segway works the same way. The farther the driver leans, the faster the wheels rotate.

## *The Gyroscope*

A spinning top keeps on spinning, without falling over, precisely because it is spinning. Spinning objects, as large as a planet or as small as a top, tend to maintain their orientation, and that fact is exploited in the gyroscope. Gyroscopes are often used for maintaining the stability of objects such as airplanes. The first gyroscopes, in the early 1800s, were strictly mechanical and utilized a rotating element reminiscent of a spinning top. By the early 1900s practical gyroscopes were used in navigation.

Many modern gyroscopes are vibrating structure gyroscopes. Such gyroscopes are used in the Segway, the iPhone 4, various video game controllers, and the image-stabilization systems of many still and video cameras. Instead of a spinning mechanism, vibrating structure gyroscopes are based on a vibrating element. And since they are present in devices like cell phones, they are obviously pretty small.

Vibrating structure gyroscopes have a close biological cousin in the form of specialized sense organs, called halteres, found on many flying insects, including flies and mosquitoes. Each halter (halteres is the plural) consists of a whiskerlike appendage with a little ball attached to the free end. The halteres are located just behind the insect's wings, one on each side, as shown in figure 14. When the insect flies, the halteres vibrate. When the insect changes direction in flight, the vibrating halteres tend to resist the change in direction. Much as a spinning top tends to maintain its orientation, a vibrating object also resists changes in direction. When an insect with halteres changes direction, the ends of the vibrating halteres that are attached to the insect's body experience forces caused by the change in direction. Those forces are detected by sensory receptors that send signals to the brain, acting as a guidance system. The vibrating

*Fig. 14.* A tipulid fly. The two halteres are clearly visible behind the wings on either side of the abdomen. (Photograph by the author)

structure gyroscope thus shares both operational principles and some applications with the halteres. There are a wide variety of different designs for human-made vibrating structure gyroscopes, and they don't necessarily bear much physical resemblance to halteres.

## Sensing the Linear Motion of the Head

The other two sensory organs in the vestibular system, called the utricle and the saccule, sense the linear, as opposed to the rotational, acceleration of the head, as well as its position. Together, the utricle and the saccule are known as the otolithic organs, for reasons discussed below. These ingenious little devices share some features with the semicircular canals but are different in crucial ways. Both the utricle and the saccule are tiny sacs containing not more than a few drops of fluid, and each also has a patch of hair cells that send signals when they are stimulated (moved) by the motion of that fluid. The patches of hair cells of the utricle and the saccule lie in planes that are perpendicular to each other. The hair cells of the utricle extend vertically like tiny blades of grass in a horizontal plane when your head is in its typical upright position. Thus, the utricle is most sensitive to motion forward and backward or left and right. The hair cells in the saccule are oriented at ninety degrees to those of the utricle, and they are most sensitive to up-and-down or forward-and-backward motion.

A critical difference between the semicircular canals and the utricle and saccule is that the latter are capable of sensing position, not just acceleration. When you tilt your head forward to look down at your feet, the utricle and saccule (along with the semicircular canals) sense the movement you made. But once you've stopped in that new position and for as long as you remain there, the semicircular canals don't sense anything. They're simply waiting to detect the next rotational motion. The utricle and saccule, because of the way they're constructed, are able to "remember" the new position the head is in. They not only trace the acceleration it took it to get there, but they keep track of where the head ends up.

In the world of the senses, there are certain problems with what are known as "maintained stimuli" that continue, unchanged, for a long time. If you stare at the same image for a long time, it can sort of imprint itself on your brain, so that when you look away, a ghost of the image remains. Most sensory receptors have ways of adapting to maintained stimuli, and so it is with the utricle and saccule.

The fluid inside the utricle and saccule is the same as that found in the semicircular canals. But the membrane that contains the hair cells inside the utricle and saccule is different from the membrane inside the semicircular canals. The latter is almost perfectly elastic in its behavior—like a rubber band. When the fluid inside the canals pushes the membrane in one direction or the other, it deforms like a rubber band. But when the fluid force stops, the membrane returns to its original shape, just as a rubber band would.

The membranes inside the utricle and saccule do not behave in a fully elastic manner. Those membranes, like the membranes inside the canals, are filled with a gelatinous substance. However, and this is the key point, the gelatinous material in the utricle and saccule also contains tiny solid particles variously called otoconia (Greek for *ear dust*) or otoliths (for *ear stones*).

Imagine the membrane in a utricle or saccule as a patch of artificial grass lying at the bottom of a stream of water. When the water flows in one direction, the blades of grass bend in that direction. When the water stops flowing, the blades of grass return to their original upright position. The blades of grass, which represent hair cells, in this case have no memory of their position. To provide the crucial memory feature, imagine that the patch of artificial grass is filled with tiny grains of sand. Now, when the water flows in one direction, the blades of grass still bend with it, and the grains of sand shift ever so slightly in the direction of the flow. When the water stops moving, the blades of grass can't return to their original positions because they are being constrained by the grains of sand. The blades of grass thus "remember" their position.

The part of the brain that is keeping track of the position of the head is thus receiving a continuous signal. "Okay, the head's tilted down toward the feet. It's still tilted down. It's still tilted down. Now it's tilting back up again . . ."

The gelatinous, particle-filled substance in these organs behaves sometimes like a fluid and sometimes more like a solid. Materials that have this sort of split personality are called Bingham fluids. Bingham fluids are said to possess a "yield strength," something that is relatively rare in the world of fluids. Imagine a classic fluid mechanics experiment, in which you trap a thin layer of fluid between two flat, horizontal pieces of glass. When you push the top layer of glass sideways while the bottom layer is stationary, the top glass will always move, even if the sideways force you apply is very slight. This is because of the slippery layer of liquid in between the layers of glass. Olive oil is more viscous (thicker) than water, but it would still behave in the same way in the sliding plate experiment: as soon as you push on the top plate, it would begin to move. But not if you used a Bingham fluid; Bingham fluids don't behave this way. An everyday example of a Bingham fluid is toothpaste, which doesn't begin to flow until a certain amount of force (the yield stress) has been built up. Bingham fluids, like toothpaste or mud, tend to be slurries. A slurry is a liquid with a high percentage of solid particles suspended in it, and that's what you find in the utricle and saccule.

With its ten specialized accelerometers, six semicircular canals for measuring rotations and four otolithic organs for tracking linear motion and position, the vestibular system keeps track of the acceleration and position of the head with superb accuracy. It is ready for just about any sort of situation the natural world can throw at it. But it wasn't designed with airplanes in mind.

## Sensory Illusions in Aviation

If you want to fly airplanes, you must trust your equipment. Pilots have to believe in the integrity of their airframe, the reliability of their engines, and the consistency of their controls. But the trust that pilots must place in their equipment goes well beyond this in at least one other crucial, life-and-death way. Pilots must trust their instruments, the sensory systems of their airplane, so profoundly that they believe them, the instruments, instead of what their own senses are telling them. This is because the very nature of flying is so foreign to our senses that we are easily fooled. In an airplane, our senses, particularly the vestibular system, tell us that the plane is doing one thing, when in reality the plane may be doing just the opposite.

Humans rely on a complex integration of three different senses for "stability," or the capability of maintaining a given desired position of our body, such as standing erect. The three stability-enabling senses are the eyes, the vestibular system, and the proprioceptive sensors in our extremities (proprioception is the topic of the next section). Imagine that you are on the pitching deck of a ship in a rough sea. What keeps you from falling down? Your balance system senses all the various rotations and linear motions your head is undergoing as the ship's deck rises and falls. Your brain processes that balance data and uses it to control your eyes, hands, and feet. Your proprioceptors in turn keep your brain constantly updated as you shift your feet and hands to keep your balance. And your vision is observing the pitching deck and the next wave that's coming, helping to prepare you. The brain somehow integrates all these inputs and coordinates an overall strategy for keeping you upright. Vision is perhaps the least important member of this triumvirate, as unlikely as that may seem. We are perfectly capable of standing upright in profound darkness, for example. When there are problems with the vestibular or the proprioceptive systems, as in Parkinson's disease patients, however, this may not be true.

The vestibular systems of airplane pilots are typically in perfect working order. The problem is that those systems weren't designed for the types of inputs they receive in a flying airplane. Especially when a pilot is not able to see, at night or in inclement weather, bad things can happen, particularly if the pilot has not learned to trust the plane's instruments.

There are so many different sensory illusions taking place in aviation that it would be hard to list them all. One such sensory illusion is sometimes called "the leans." If a plane has been making a very gradual and prolonged turn, and then the pilot suddenly levels the wings, his senses are likely to tell him that the plane is in fact now turning in the opposite direction. As a result he is likely to "lean" back into a turn in the original direction. His own senses are telling him that is what he needs to do to keep the aircraft level. A long, gradual turn like this fools the senses in two ways. First, the semicircular canals are designed to detect rotation of the head, but if the head is rotating very slowly, the canals can't sense it. Such motion falls below the limits of detection of the semicircular canals, which can't detect angular accelerations of less than about two degrees per second. In everyday life on the ground, your head rarely if ever rotates that slowly,[3] so it's generally not a problem—unless you are a pilot.

When I first learned about this two-degrees-per-second detection limit, I decided to do some experiments. The first thing I tried was closing my eyes and then turning my head to one side as slowly as I could. Before that, I closed my eyes and then turned my head at more normal speeds. I found it was relatively easy to sense where my head ended up. For example, while looking at my computer screen, if I close my eyes and then turn my head toward the door, I find that when I open my eyes I am indeed

---

[3]This works out to a complete rotation in no less than three minutes.

looking straight out the door. I tried the same experiment, only this time turning my head as slowly as I could bear, with my eyes closed. The experiment didn't work out as I'd hoped. It's hard to fool your brain this way, since there are *three* sensory systems working in concert here. By closing my eyes I've taken away vision, and I'm trying to fool the vestibular system, but the proprioceptors, particularly those in the neck, are still on task, and I can feel my neck muscles changing as my head is rotating ever so slowly. So I know my head is moving. But during this little ad hoc experiment, I nonetheless experienced a strange floating sensation that's difficult to describe, so I decided to attempt something a little more sophisticated.

I coerced one of my graduate students to sit down on a lab stool that has a swiveling seat and a footrest. I blindfolded her and asked her to determine how far I rotated the stool. At first, I tried to rotate it at roughly the speed of normal head movement. I told her that I was going to rotate her either 0, 30, 60, 90, 120, 150, or 180 degrees—anywhere from zero to one-half of a full rotation. Before the blindfold went on, we practiced the drill, and I placed landmarks on the wall at eye level at each of the thirty-degree increments. Blindfolded, she did an excellent job of sensing how far the stool had been rotated, as long as I was rotating it at normal speeds. In the second part of the experiment, I rotated the stool extremely slowly, at two degrees per second or less. This meant that a 180-degree rotation would require at least a minute and a half. So I told her I was going to rotate the stool very slowly for one and a half minutes and asked her to determine how far she'd been rotated. She couldn't do it. When we switched roles, I was similarly unable to detect how far the stool had been rotated.

This result shouldn't be surprising. All of our senses exhibit detection limits. The average person cannot hear very faint sounds. The threshold of audibility is often found to occur at a sound pressure level of about zero decibels at a frequency of 3,000 Hz. Similar detection limits exist for odors, seeing in dim light, and so on.

Returning to our airplane pilot who's trying to deal with the leans: I mentioned that there are two problems with detecting a long, slow turn. The first is the one discussed just above, that the semicircular canals can-

mundane task, carefully avoided drinking from any but the most robust beverage containers.

At the beginning of part 2 I introduced a scheme for categorizing the body's sensory receptors as photoreceptors, chemoreceptors, thermoreceptors, mechanoreceptors, and so on. In this scheme, the sensors responsible for proprioception are mechanoreceptors, since they monitor things like the length and tension of muscles or the position of a joint.

An earlier scheme for classifying sensory receptors divides everything into three categories: interoceptors, exteroceptors, and proprioceptors. This system was proposed by the British Nobel Prize–winning neurologist and physiologist Charles Scott Sherrington in 1906. Interoceptors monitor strictly internal bodily events. The receptors that tell you your stomach is full and those that measure the chemistry of your blood are interoceptors. Exteroceptors include the rods and cones of the eye, the olfactory receptors, and the touch receptors. These all respond, as the name suggests, to external stimuli. Proprioceptors respond to changes in body position, as exemplified by the receptors in the muscles and joints. Sherrington's classification of receptors is less popular today than the one we've utilized throughout this book, but it has its merits. Both systems suffer from problems of overlap. In Sherrington's scheme, the balance system would be part of proprioception, responding as it does to the position of the head. But the body's balance system responds to external stimuli as well, such as the motion of the pitching deck of a ship. In similar fashion, vision, which generally deals with external stimuli, also plays a proprioceptive role in monitoring body position and motion.

We concern ourselves here with the proprioceptors that monitor muscle and joint position and motion. Since Ian Waterman lost his proprioceptive and tactile abilities, but not the ability to sense temperature or pain, we know that different kinds of receptors are responsible for sensing these various stimuli.

Muscle tissue, like skin, is filled with nerve endings. The roles of all these receptors are not completely understood. Some are pain receptors, while others likely monitor chemical changes in muscle fluids. Proprioception in muscles and tendons is handled by receptors with names like muscle spindles and Golgi tendon organs. Muscle spindles are receptors that are activated by stretching. When a muscle stretches, these receptors stretch along with it, mechanically opening channels in the receptors that result in voltage changes. Muscle spindles come in a wide range of nuanced varieties, so that some spindles essentially act as measuring sticks, keeping track of how long a given muscle is at any instant; other spindles, sensitive to the rate at which a muscle's length is changing, act more like muscle accelerometers.

Muscle spindles do a great job of monitoring muscle length and its rate of change. Hold your arm out straight horizontally. Now bend it at the elbow, moving your hand back toward your head. The changes your biceps undergoes during this motion are tracked by muscle spindles. Now repeat the motion while holding an object weighing a pound or two in your hand. What you sense, obviously, is quite different. But your muscles, in terms of their length and rate of change of length, are doing the same thing in both exercises, with and without the weight. The difference is the tension in the muscles, and that is sensed by receptors in the tendons, not in the muscles themselves. Here's another exercise. Grasp an immovable object, such as the knob of a closed door, with your hand. With your arm fully extended, pull on the knob (without twisting it). What you sense is tension in the muscles of your arm. Muscle spindles are relatively insensitive to this action, since in this exercise your muscles are undergoing very little change in length. Once again, sensors in the tendons are doing the job.

Golgi tendon organs are located in tendons, near the point at which the tendons connect to muscle. These sensors have been shown to be extremely sensitive to muscle tension and help us to control muscle tension when, for example, we want to avoid damaging delicate objects such as a flimsy paper cup.

In the joints themselves there are a variety of different kinds of receptors that monitor position and movement. These joint receptors play an important role in proprioception, but not as important as was once believed. For example, a person who has had joint replacement surgery can still sense the position of that joint, even though the joint receptors are no longer there. Receptors in the muscles, tendons, and joints thus combine to give us proprioception—the knowledge of where our body parts are and what they are doing. Walking, running, playing sports, typing, writing, eating, bathing, and tying your shoelaces—the list of everyday human activities that rely on proprioception is as long as you want to make it.

# Part 3 • Perception

Stimulus and sensation exist entirely in the here and now. The sound waves bouncing off your eardrums, the light waves focused through your eyeballs and striking your retinas, the molecules entering your nose or settling onto your tongue, the accelerations of your head or the positions of your limbs—these stimuli are fleeting. Likewise, the sensations they cause, the electrochemical reactions of our sensory instrumentation, exist only for the briefest of moments.

Perception, in sharp contrast, is never entirely in the present. It depends not just on the present-tense stimuli being processed by the sensory receptors, but also on all the past-tense experiences to which a person has ever been exposed. Presented with the same stimuli, the perceptions of two individuals can be vastly different. Some of those differences are shaped by differences in our sensory instruments, by old versus young ears, for example. But crucial variations in perception arise out of our experiences. Three people with normal hearing listening to Louis Armstrong's famous 1927 recording of "Potato Head Blues" might all enjoy the music. However, one of the three might be unable to identify the lead instrument in the recording. The second might recognize it as a cornet, and the third might be able to identify the distinctive playing of Satchmo himself. Each person brings his or her own perceptual history to the experience of listening to the music.

Perception is cultural, too. The sense of smell, the sensory hardware, is the same in an American as it is in an African or an Asian. But the perception of odors varies from one culture to the next. The durian is sometimes referred to as the "king of fruits" in Vietnam. Roughly the size and shape of a rugby ball, it has a thick, green, thorn-covered skin. Even with the skin intact, the odor of the fruit is quite strong. To many Vietnamese, that odor is heavenly. People from other cultures often find it unpleasant if not downright nauseating. Similar cultural differences relate to tastes, sounds, and even sights.

It's hard to know when sensing stops and perceiving begins, so intertwined are these two processes. According to Merriam-Webster, perception is "awareness of the elements of the environment through physical sensation." The *American Heritage Dictionary* calls it "recognition and interpretation of sensory stimuli based chiefly on memory." The key word in the first definition is "awareness." It is not so unusual for someone with fully functioning eyes, for example, to be partially or completely unaware of the stimuli that are being delivered to the brain by that hardware. The reverse is true as well. A patient with macular degeneration, a disease of the retina, is generally perfectly capable of perceiving visual sensations. The problem is that the hardware of his eye can no longer do the job it once did. Beyond all that, awareness, and thus perception, as the Louis Armstrong example shows, is subtle and infinitely nuanced.

Perception is all about the brain, and the human brain is a magnificent, complex organ. It is generally considered to be without serious rival in the animal kingdom. While our brains may be superior, we certainly can't say that about our sensory abilities. There are animals who can see better (with more acuity, from greater distances, in the dark), hear better (over a wider frequency range, at lower sound pressures), or detect smells better than we do. There are even animals that can sense stimuli, such as infrared or ultraviolet radiation, or magnetic or electrical fields, of which we humans are completely unaware through our natural sensations. So what do we have that they don't have? A brain, as the Wizard of Oz once pointed out to the Scarecrow. To be fair, all those other animals have brains, too. It's just that ours is, well, a lot better. And having a better brain can make up for relative shortcomings in the sensory hardware. In the Darwinian struggle to survive, the strategy that has evolved for Homo sapiens seems to center on improvement of the brain, which has allowed us to dominate stronger, faster, and better-sensing animals.

# Chapter 8 • Remembering the Present

## Measuring Perception

Humankind, as noted throughout this book, has been pretty successful at perfecting instruments that sense things. What about instruments that perform perception? Given an old jazz recording of fuzzy origins, is it possible to determine whether, for example, it was performed by Louis Armstrong? Computer analysis of such a recording can indeed verify the fingerprint of Armstrong's playing, separating him from his many imitators.

There are other examples of machines that perceive. Let's talk about cars. Those in the business of designing and engineering automobiles often speak of the esoteric concept of "ride quality." Ride quality refers to the feelings one gets when riding in a car. These feelings often make the difference between buying and not buying a car based on a test drive. First impressions are important in dating, in job interviews—and in new-car shopping. Probably nothing sells cars better than how they look on the showroom floor, but ride quality during that first test drive is also crucial.

The sensations involved include vision, touch, balance, proprioception, hearing, and smell. How do automakers evaluate ride quality as a new car model is being developed? The traditional way is the road test, wherein data are gathered by an experienced professional driver equipped with finely tuned senses, including what is sometimes referred to as a "sandpaper butt." The driver needs to be able to sense the various bumps and vibrations received through his posterior, transmitted from the road, the engine compartment, elsewhere via the car's frame, and on through the seat. The role of a good road tester in evaluating ride quality for an automaker is somewhat like that of the taste tester in a food factory, or someone who can expertly smell and evaluate perfumes. All are charged with perceiving various qualities in their company's products that are difficult or impossible to measure with instruments.

In the case of ride quality, there are some pretty sophisticated products that attempt to replace the test driver through a combination of automated sensation and perception. First, key sensory information is gathered by thoroughly instrumenting a car. Accelerometers are mounted on the seat and elsewhere, and microphones are positioned where the ears of the drivers and passengers would be. Once the car is fully instrumented, it is driven through a standardized driving cycle. The noise and vibration of the engine are recorded. When an automatic transmission shifts gears, the slight jerk does not escape the notice of the instruments. Similar inputs from steering, braking, accelerating, and traveling over bumps and potholes are collected.

Those data are then compared, via a sophisticated computer program, to the subjective responses, the perceptions, of hundreds of persons who have ridden in similarly instrumented cars and later answered detailed surveys about how they perceived a great many different aspects of the experience.

The automaker can do various things with these results, including calculating a single number for the overall ride quality of the car. That number can be compared with other cars' numbers or referenced later after the car has been further developed. Combining the objective instrumental measurements (sensations) with the subjective responses of real passengers (perceptions) is analogous to how perception works in human beings. Perception is a complex blend of what's happening now with what happened before. To perceive is, somehow, to remember the present.

# Dr. P.

In his book *Genome*, Matt Ridley notes that early in our struggle to understand the roles of individual genes, we often tended to describe this gene or that not in terms of its purpose, but instead in terms of the diseases it can cause. Ridley concludes that defining genes by the diseases they cause is ridiculous. It would be like defining organs in the body that way; for example, a malfunction of the pancreas can cause diabetes. As our knowledge of the human genome has grown, this trend has begun to change. You still hear news stories about the "obesity gene" or the "alcoholism gene," but the more constructive idea of genes as the builders of the body, the keys that fit into the locks of creation, seems to have taken hold.

In a somewhat analogous fashion, much of what we know about perception comes from studies that have been done on persons with perceptual problems. We study the abnormal to infer the behavior of the normal. Take the strange case of Dr. P.

Dr. P. was a gifted musician and music teacher. Late in life, he developed a rare neurological condition that manifested itself as a bizarre vision problem. An ophthalmologist concluded that there was nothing wrong with his eyes, but rather that the problem lay in the part of his brain that processed data from his eyes, the visual cortex. The visual cortex is the single largest region in the brain, containing about ten times as many nerve cells as are devoted to hearing. Although the process by which the visual cortex processes the signals it receives from the retina is enormously complex, a great deal is known about it.

The ophthalmologist recommended that Dr. P. visit a neurologist. That neurologist was Oliver Sacks, and Dr. P. became the title character in Sacks's book *The Man Who Mistook His Wife for a Hat*. Dr. P. had begun to have difficulty recognizing people. His students were among the first to notice this. He would sometimes not recognize a student until the student had spoken, at which point the student's voice would provide an instant identification. Sacks's description of his examination of Dr. P. is riveting reading. It describes a man whose condition probably doesn't resemble anything you've ever encountered—unless you're a neurologist.[1]

In response to routine neurological tests, Dr. P. was able to visually identify standard shapes such as a cube and a dodecahedron (a solid with twelve pentagonal faces). He was also able to identify playing cards randomly selected from a deck, including the face cards and the joker. The ophthalmologist was right; there was nothing wrong with Dr. P.'s eyes. But when Dr. P. was shown a series of photographs of family, friends, colleagues, and even of himself, he recognized almost no one, not even himself. Keying in on the famous hair and mustache, he did recognize a photo of Albert Einstein. But lacking obvious, almost cartoonlike features, the types of features that had allowed him to identify the jack, queen, king, and joker in the deck of cards, he was lost.

Sacks handed him a rose. Dr. P. noted that it was "about six inches in length. A convoluted red form with a linear green attachment." Encouraged to move beyond description to identification, Dr. P. responded that it wasn't easy. Sacks suggested that he smell it. "Beautiful!" said Dr. P. "An early rose. What a heavenly smell!"

The clinical name for a loss of the ability to recognize familiar faces is prosopagnosia, or face blindness. It is a specific form of agnosia, which is a general difficulty in recognizing familiar objects with the senses. Sacks compares Dr. P.'s affliction to the way a computer interprets images. Dr. P. constructed his visual world, as best he could, by means of "key features," such as Einstein's wild hair. But he was unable to grasp the totality of an image, such as a face or the rose. A face was just a set of features. If those features were distinct or unusual enough, as with Einstein, he could make the leap to an identification. But Dr. P. could not see or relate to the *wholeness* of a face or an object, only to those features he could extract. Dr. P.'s condition was brought on by a massive tumor in the visual cortex of the

---

[1] Sacks himself suffers from a lifelong difficulty in recognizing people's faces— even his own.

brain. Although his vision problems worsened, Dr. P. performed and taught music until the very end of his life.

Most of us are at least vaguely aware that different parts of the brain are responsible for different functions, things like vision, hearing, muscle control, emotions, and so on. But the integrative aspects of perception, the idea that certain parts of the brain might process the details of vision, the ones Dr. P. could perceive, whereas other parts put together or integrate those details, which Dr. P. couldn't do, is perhaps less appreciated. Nonetheless, it is fundamental to an understanding of perception.

## Face Recognition

How many passwords and personal identification numbers (PINs) do you have? If you're like me, the answer is "too many." We have passwords to log onto computer systems, e-mail accounts, online bank accounts, and online shopping sites. And we use PINs to gain access to ATMs and to use a debit card. Wouldn't it be nice if all of these electronic systems could verify our identity without our having to keep track of all those passwords and PINs? In the good old days, when you went to the bank, the teller recognized you right away and was happy to do business with you without any further identification. Why can't a computer recognize us the same way? Face recognition may be such a mundane perceptual task that most of us don't give it a second thought, but it is anything but mundane for today's computers and software.

The integrative aspects of perception, as exemplified by Dr. P.'s story, bring to mind the drive to develop automated face recognition technology. Most people can recognize hundreds if not thousands of faces. When I run into a former student (I've had thousands) I usually recognize the student's face, although connecting that face to a name, especially in time to offer a personal greeting, is a much more perilous affair.

Our ability to recognize faces lies on a spectrum described by a bell curve similar to that of an IQ test. At the upper end of the curve, there are the "super-recognizers," people with an uncanny, almost magical, ability to recognize faces. My oldest stepdaughter is a super-recognizer. She can pass someone in a crowd at the airport and remember him as the waiter who served her a restaurant meal a few years back. At the lower end of the curve are the severely face blind, who are sometimes unable to recognize the faces of family members or even of themselves. This condition can be acquired, as in the case of Dr. P., or congenital, as in the case of Oliver Sacks, the man who wrote about Dr. P.

The difference between a super-recognizer and a severely face-blind person would be the same, in terms of face recognition, as the difference in intelligence between individuals with IQs of 150 and 50. As with IQ and other things described by a bell curve, most of us are bunched in the middle. Face blindness has begun to be studied in detail only in the past few decades. It is now believed that at least 2 percent of the U.S. population, more than 6 million people, are severely face blind. One of the leading research centers on face blindness is directed by Ken Nakayama of Harvard, who is himself a mild sufferer. On his Web site, faceblind.org, you can take several tests to determine your own abilities. I took one of the tests and scored below average, which surprised me a little.

On the test, photos of the faces of celebrities are displayed one at a time. The images are cropped so that only the face is displayed—no hair, neck, or shoulders are visible. After looking at a photo, you type in the person's name or something about him or her. For example, if you recognized Russell Crowe's face but couldn't remember his name, you could just type in "star of *Gladiator*," and that would be a correct answer. This is a test of your ability to identify familiar faces, not your ability to remember names. If the image is of a celebrity you are unfamiliar with, that doesn't count against you. I know the name Justin Bieber, for example, but I have no idea what he looks like. My inability to recognize his photo would not count against me.

The very first photo on the test was of a famous actress I'm quite familiar with, but I didn't recognize her from her face alone. I responded "I don't know," and her name then appeared on the screen. I thought to myself,

"That's not what she looks like." But then I realized that I must identify this particular actress by her distinctive hair. From then on, I tried hard to "think beyond the hair," but it was difficult. I was unable to identify several other photos of people I'm quite familiar with. I coerced my wife into taking the test, and as I had predicted, she did extremely well. She failed to identify only one out of thirty photos, and the one she missed was someone she barely knows.

What is it about human faces that make them so recognizable? A face is both an entire thing and a collection of details. As the case of Dr. P. shows, there's more to recognizing a face than just picking up on the details. We recognize the *entire* face, along with the details. Even when the details of one of my former student's faces have changed over time—a once bearded face may now be clean-shaven, for example—I still generally recognize the face, although my score on the faceblind.org test has caused me to reevaluate my abilities in this regard.

Research into computer-based face recognition dates back to the 1960s. Ever since then, researchers have been struggling to program computers to be able to identify the photographic or video image of a face by comparing it to a database of other faces. In the wake of September 11, 2001, face-recognition research received a boost as part of increased efforts to identify terrorist suspects at airports and elsewhere. By that time, computers had become fast enough to make it at least theoretically feasible for them to survey a constantly changing collection of faces in a crowd, compare them to database images, and provide matches.

Until relatively recently, face-recognition software required a fairly controlled type of image, such as a frontal photograph of a face, taken in good lighting. Such photos could be compared to a database of driver's license photos, for example, to find a match. Changes in lighting, the orientation of the face, facial expressions, and so on, could confound such systems rather easily, leading to both missed identifications and misidentifications.

The ability to perform "3D" face recognition has improved impressively in recent years. Numerous companies, such as L-1 ID (www.l1id.com), provide commercial services in this area. A face in a crowd rarely presents itself to a video surveillance camera head-on, like a driver's license photo. A 3D software system first has to identify the faces in an image, by sorting the faces out from the rest of the image. Once an individual face image has been brought into focus, various measurements are taken. Every face is a collection of quantifiable landmarks, like the topography of a mountain range. The distance between eyes, the width of the nose, and so on, can be measured. When the image of the face is at an angle to the camera, such measurements must be calculated from the geometry of the image, essentially transforming an image taken at any angle into a two-dimensional frontal image. Matching the resulting 2D image to the 2D images in a photographic database is a matter of comparing as many as eighty of these topographical measurements.

But such comparisons are not always sufficient. A relatively recent advance in face recognition utilizes a technology somewhat analogous to fingerprints. A "skinprint" is a detailed photographic image of a patch of facial skin, which is converted into a set of mathematical parameters based on lines, wrinkles, pore structure, and the like. Skinprints are said to be able to differentiate identical twins, something not possible using only topographical measurements. Because skinprinting is computationally intensive, it is typically used after topographical measurements have created a relatively short list of potential matches.

Lots of things create challenges for computerized face-recognition systems. Looking to escape detection? Wear glasses, or better still, sunglasses. Grow your hair long, and let it hang over your face as much as possible. Wear a hat. Stay out of well-lit areas, and don't stand in one place for too long. In general, just be as shady as possible.

Face recognition is not the only visual perception task we humans have attempted to computerize and thus automate. A repetitive visual perception task in the medical field, such as interpreting a mammogram, is another good example. Such a task is relatively routine, but the consequences of getting it wrong are significant. As with face recognition, however, developing a system that truly automates such a visual perception task is extremely challenging.

How does a computerized system for face recognition, or some other visual perception task, compare with the system your brain utilizes? Computers and software have different strengths and weaknesses compared to the human brain. A distinct advantage for computers is the number of faces they can store in a database and thus "remember." Humans are far more limited in this respect. In contrast, the supremely nuanced ability we have to recognize faces based on so many different factors, some of which probably haven't even been identified yet, gives us a significant advantage over the computer. For one thing, humans integrate inputs from multiple senses, not just vision, to identify faces in real, face-to-face life. Obviously, this does not apply when we identify faces from photographic images. We also use context to help recognize faces. I do a better job recognizing students' faces at school than if I bump into a student somewhere else. Oliver Sacks tells of a time when he did not even recognize his own assistant of many years when he ran into her someplace outside their normal routine.

It could be that face recognition is such an important perceptual skill for humans that it has evolved into a unique process in the brain, separate from other visual object identification tasks. Evidence for a special process includes our apparent ability to remember human faces better than other types of objects. Another piece of evidence: the face blind frequently, unlike Dr. P., have no other serious visual symptoms. That is, they can visually identify many familiar objects other than faces. Results of brain imaging studies suggest, however, that face recognition may share much in common with other types of object identification.

Whether face recognition by humans depends primarily on a holistic evaluation, as seemed clear for Dr. P., or on the integration of data from individual features, such as big ears or a crooked smile, has been the subject of much study. In the introduction to this book, I passed along Betty Edwards's advice to practice drawing faces by turning photographs upside down and then sketching them. That way, you are forced to draw what you really, objectively see. Your brain has difficulty doing a lot of perceptual postprocessing on an upside-down facial image. Try looking at some upside-down photographs of the faces of famous people. You will find that they are much more difficult to identify than when they are right-side-up. This is seen as evidence that faces are perceived holistically rather than as a collection of features. There is an interesting corollary to the upside-down-face phenomenon. When individual facial features, such as the mouth or the eyes, are flipped upside-down and reinserted into a right-side-up face, strange perceptual things happen. A right-side-up face with an upside-down mouth looks positively grotesque.[2] The same image, when inverted so that the face is now upside down and the mouth right side up, looks so relatively normal that many people don't even recognize that the mouth is inverted relative to the face. This is known as the "Thatcher illusion," first studied by Peter Thompson in 1980.

Lots of other aspects of human face recognition have been studied. Consider caricatures. When we see a political cartoon of Barack Obama, we recognize his face just as quickly, if not more so, as we would from a photograph. A computer algorithm, designed to recognize photographic facial images as described earlier in this section, could never hope to recognize the caricature. The topographical features have been wildly distorted, and there is no possibility of doing a skinprint. Why are caricatures so easily recognized by humans? This could be an argument for the importance of features in face recognition. Caricatures of Obama nearly always feature huge ears, enlarged front teeth, and an exaggerated square, jutting jaw. Even simple line drawings containing these caricatured features are instantly identifiable. It seems that face recognition by humans depends on both holistic and feature-based aspects.

But we do more than just recognize faces. A face also provides important clues about its owner's emotional state, and humans are very good at ferreting out whether someone is happy, sad, afraid, angry, or excited. People are naturally so good at this that serious card players must practice their poker faces in order to avoid giving away valuable clues, known in the business as "tells," about the cards they are holding. Some face-blind individuals have demonstrated the ability to discern someone's emotional

[2]A Google image search of "upside down faces" reveals quite a gallery of these.

state based on a facial image. So we must add yet another layer of complexity to our perception of the human face.

The big picture versus the details. A holistic, feature-integration approach to face recognition versus a detail-oriented feature-extraction way of looking at things. These two approaches to perception show up over and over again.

Even at the level of the whole person, some of us seem more suited to one approach than the other. When I began graduate school, my research adviser told me that the world of research was made up of two kinds of people: helicopter pilots and bloodhounds. Imagine you're searching for a missing person in a vast wilderness. Finding the person is like the goal of a research project. Searching for him or her is the research process itself. Which would you rather be, my adviser asked me, a helicopter pilot or a bloodhound? The helicopter pilot can cover territory very rapidly, but the chances that he will fly right over the missing person without realizing it are great. The bloodhound covers territory much more slowly but is almost certain to find the missing person if he happens to be close by. Helicopter pilot—big picture; bloodhound—detail-oriented. My adviser told me the best research partnerships pair a helicopter pilot with a bloodhound. I think he was exactly right. I am definitely a helicopter pilot when it comes to research. Over the years, the graduate students I've had the most success with were dedicated bloodhounds.

## The Hardware of the Brain

Thinking about the brain is a little like thinking about the universe. Both are complex, wondrous creations, and both involve lots of really big numbers—so big they are hard to imagine. The average human brain, for example, contains about 100 billion neurons or nerve cells. That's about fifteen times as many nerve cells in one brain as there are living human beings on Earth.

The brain is also often compared to a computer, and in some ways that comparison is useful. A computer, straight out of the box, has the ability to do lots of different things, depending on how it is programmed. Ten identical computers, removed from their boxes and given to ten different people, will likely have very different capabilities one year later, depending on the software each individual installs on his computer and the ways in which each machine is customized. It's nature versus nurture. All ten computers have an identical nature, but they become very different machines based on how they are nurtured by their owners.

When brains come "out of the box," that is, when we are born, the nurturing process, the way we are raised by our parents, taught in school, and influenced by our friends, our cultural environment, and so on, has everything to do with how our brains develop and who we become.[3]

Academics generally believe that their own research specialty is the most interesting thing that anyone could ever study. You almost have to believe that to make it through all those years of graduate school and wind up with a Ph.D. So when Jeannette Norden stood up in front of me and an audience of hundreds of professors from every academic field imaginable and told us that we were all wrong, that whatever our academic specialty was, it was anything but the most interesting thing going, well, it certainly got our attention. We were all wrong, Norden said, because there is nothing that anyone could possibly study that is more interesting than the human brain. Norden is a professor at the Vanderbilt University Medical Center, a dedicated and innovative teacher, and a spellbinding storyteller. If you didn't believe that the brain was the most interesting thing you could possibly study before hearing Jeannette Norden lecture, I'd be willing to bet that you would think so afterward.

The human nervous system is divided into two parts, the central nervous system, which consists of the brain and the spinal cord, and the peripheral nervous system, an enormous collection of nerves that branch out to nearly every part of the body. The retina is often classified as part of the central nervous system, since the retina and the optic nerve arise as outgrowths of the developing brain.

---

[3]The nurturing process inside the womb is important too.

There are hundreds of billions of cells in the nervous system. Nerve cells process and transmit information through electrical and chemical means and come in all shapes and sizes. But they have a few things in common, as shown in figure 15. At one end are generally found a series of branches, called dendrites (from the Greek for *branch*). In a really large, complex nerve cell, there could be thousands of branches. This gives the cell a strong resemblance to a fine network of plant roots. The branches receive information, in the form of flows of electrical charge, through contacts, called synapses, with other nerve cells or with muscles or other sensory stimuli, such as connections to the sensory receptors discussed in part 2.

Both *neuron* and *synapse* are terms coined by Charles Scott Sherrington, whom we met in chapter 7 when we discussed his classification scheme for sensory receptors. *Synapse* comes from the Greek *synaptein,* meaning to fasten or bind together. The synapses between nerve cells can involve either chemical reactions or more direct electrical connections, with the goal of transferring electrical charge from one nerve cell to the next. The signals are transferred the length of the cell—some can be up to a meter

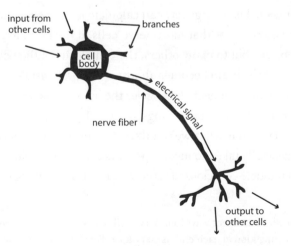

*Fig. 15.* A typical neuron or nerve cell.

long—where they are passed on to other nerve cells through another set of synapses.

Many sensory receptors are nerve cells. Touch and temperature receptors, for example, are nerve cells that are directly sensitive to their respective stimuli. They create an electrical signal at one end and pass that signal off to another neuron at the other end. Other sensory receptors, such as the hair cells in the cochlea, are not considered nerve cells. They process sensory stimuli and pass them on to nerve cells, which pass them to other nerve cells, and so on. Electrical signals from sensory receptors thus generally flow in just one direction: from the extremities toward the brain.

It is tempting, then, to think of the perception process as a linear one, in which electrical signals move up the line from a stimulus to a series of relay stations in the brain, where various aspects of the signals are extracted and further processed. This view of perception, which held sway for many years, has gradually been replaced by the realization that the process is both more complex and highly nonlinear.

That the process is complex is not hard to believe. If a brain contains 100 billion nerve cells, and if each one is connected to only one other nerve cell at each end, the number of possible ways in which they could be interconnected is so large few can calculate it, or even contemplate it.[4] And when you consider that most nerve cells in the brain are connected not to one or two but to many others, the complexity of the possible interconnections in the brain becomes almost incomprehensible.

This level of interconnectedness gives the brain phenomenal computing power, allowing it to do many things at once. The subject of multitasking is taken up later, but what we're talking about here is different. Multitasking is generally taken to mean a person's ability, or inability in my case, to perform multiple conscious tasks at the same time. Talking on the phone

---

[4]The number of different ways four nerve cells can be interconnected, with either zero or one connection at each end, is sixty-four. For seven nerve cells, the number is more than 2 million, and for forty-five, the number is larger than $2 \times 10^{298}$.

while typing on a keyboard, for example. While different people seem to have different capacities for that sort of thing, what is generally true is that your brain is constantly processing all kinds of things, without your having to concentrate on most of them. Sensory information continuously pours into the brain from all those receptors. While I have to think hard about the sentence I'm writing, my ears, in parallel and without my having to concentrate on it, are constantly delivering new audio information to my brain, which my brain is updating hundreds of times a second. If the phone or the doorbell rang, my brain would send a signal, interrupting my conscious thought process. The other senses, likewise, are similarly occupied. This is true "parallel processing." The brain really is doing all of these things at once, in contrast to a computer with a single CPU, which seems to do things in parallel only because it does things so fast. In reality it is just jumping from one thing to the next, while only working, at any instant, on a single task.

## Mind and Brain

When a computer doesn't do what we want, we sometimes speak of the blasted thing as having "a mind of its own." A computer doesn't have a mind of its own, does it? Do humans have minds of their own? What is a "mind" anyway, and how is it related to the brain? These are philosophical questions, but they are related to perception. Throughout much of the history of philosophy, a strong current of thinking has maintained that mind and brain are separate entities. It is possible to think quite deeply about the brain and what it does without ever considering neurons, synapses, and the like. The more we learn about the brain and how it works, however, the less likely it seems that the mind is a separate, nonphysical thing, mysteriously connected to the physical brain.

In the world of computers, we speak of hardware and software. The hardware of a computer consists of the things you can touch, such as the monitor, the keyboard, and the mouse. Inside the computer, there is more hardware: a power supply, one or more disk drives (although these are disappearing), some random access memory (RAM) chips, and the most important hardware of all, the CPU, or central processing unit. The CPU, these days, is a silicon-chip device with billions of individual circuit elements, such as transistors, built into it. These circuit elements are roughly analogous to the nerve cells in the brain. In the early days of computers, back in the 1950s, computers had to be physically rewired to perform different tasks. The modern chip-based CPU is far more versatile. The internal circuitry of a CPU is automatically rewired by each different software program it is asked to execute. The software changes the ways in which the individual circuit elements in the CPU interact.

The brain consists of hardware and software, too. The 100 billion nerve cells in the brain are hardware. Nerve cells are connected to one another, but those connections can and do change. Making new connections among neurons and changing old connections represent the software aspects of the brain. The process by which these connections change throughout our lives is known as neural plasticity, one of the most wonderful, complex, and important aspects of how the brain works.

Our brains are faced with a formidable task when it comes to making any sort of sense out of anything, given the unrelenting, overwhelming nature of the sensory stimuli constantly bombarding us. It is no wonder infants are so helpless. It takes years to figure out how to filter, sort, categorize, and otherwise understand all those signals coming in. So much of childhood is devoted to learning how to perceive the world around us. Infants' toys such as blocks and mobiles hone their vision and sense of touch. Parents speak endlessly to their uncomprehending infants, knowing they will not be uncomprehending for long. Learning to walk is mostly about muscle control, another monumental task for the brain, but it also needs the senses of proprioception and balance.

There are many ways to think about how sensory information is processed in the brain. For example, there are the concepts of "bottom up" and "top down" processing. These terms refer to the genetically older parts of the brain that filter and separate streams of sensory data from the bottom up, while simultaneously the genetically newer portions of the brain process and interpret information from the top down. Then there are the familiar left- and right-hemisphere differences in the brain. In terms of the senses, the right hemisphere is continuously providing a sensory snapshot of each succeeding moment. The left brain is charged with responding to those sensory stimuli, the moment-by-moment patterns created by the right brain, and, among other things, placing value judgments on them, such as good versus bad or like versus dislike. Finally, there is the hugely important concept of neural plasticity, the seemingly miraculous ability of the brain to rewire itself as it learns or in response to physical damage. Our nerve cells, unlike almost all other cells in the body, do not reproduce themselves to a very great extent. That's the bad news. When you lose a nerve cell, it's typically gone forever. The good news is that the connections among our nerve cells can and do change. Neural plasticity helps explain, among many other things, why blind persons often have such acute hearing and why stroke victims frequently recover, over a period of months or years, many of the abilities they lost immediately following their stroke.

## Specialization within the Brain

A few weeks after Oliver Sacks's article about face blindness appeared in *The New Yorker,* the magazine published a letter from a woman who had suffered a minor stroke during childbirth. As a result, she had developed a condition called phonagnosia, which might be considered the hearing-related equivalent of prosopagnosia, or face blindness. When someone telephones her, this woman perceives all sorts of details or features about the voice, such as the gender and age of the speaker, and she has no trouble understanding what the person is saying. But she cannot determine whether the voice belongs to someone she knows, even if that person is her husband or another family member or a close friend. Her stroke evidently damaged the part of her brain that catalogs the voices of familiar persons.

The notion that different areas in the brain are responsible for different functions dates back at least several hundred years. The Austrian anatomist Franz Joseph Gall (1758–1828) theorized that the brain was divided into twenty-seven separate regions. Each region was responsible for a different faculty, some perceptual, some cognitive, such as memory or language, and some related to things like friendship or pride. Gall's work got him in hot water with the Catholic Church and was not particularly well accepted by the scientific establishment either, since most of his ideas lacked experimental evidence.

Gall may have gotten the details wrong, but evidence for specialization within the brain was there. The work of French physician Pierre Paul Broca in the 1860s came to represent a milestone in brain research. Broca discovered what became known as "Broca's area" near the front of the brain, a region vital to speech production. He made his discovery by dissecting the brains of patients who had suffered from *aphasia*, a general term referring to any sort of language impairment. Broca's work related to patients who had difficulty speaking. His autopsies revealed lesions in a specific region of the brain and led him to the conclusion that that region was responsible for speech production.

After Broca, researchers discovered numerous other areas in the brain with specific abilities. In 1874, the German neurologist Carl Wernicke, studying patients with brain injuries, identified a region crucial to language comprehension that is now known as Wernicke's area. Patients with injuries in this region of the brain often suffer from an inability to comprehend spoken language. Such patients retain the ability to speak, although their speech, while sounding normal, is often filled with errors.

Language impairment can take a surprising number of forms. The author Howard Engel, having suffered a stroke without knowing it, was startled to find that he could not read the newspaper he had just collected from his front porch. The front page appeared to be covered with unintelligible squiggles. And yet he could form thoughts in his head and then write them legibly on paper with a pen or pencil. In other words, he retained the ability to *write*, without being able to *read* the words he had just written. To be sure, this is a rare condition, but other documented cases do exist. Since his stroke, Engel has written several more books, including a 2007 memoir entitled *The Man Who Forgot How to Read.*

After Broca and Wernicke, the idea that different parts of the brain are responsible for different faculties became accepted, and by the early 1900s, detailed functional maps of the brain had been developed, as shown in figure 16.

Functional maps of the brain, such as Korbinian Brodmann's in figure 16, were hardly the last word on this subject. Modern brain research continues to refine our knowledge of specialization within the brain.

Techniques such as functional magnetic resonance imaging (fMRI) have confirmed the notion of specialization in the brain and also refined and expanded that knowledge. Using fMRI, changes in blood flow in the brain can be mapped in real time as a patient performs different mental tasks, such as reading, speaking, or listening to music. When nerve cells are active, when they are passing electrical signals to one another, there are changes in the chemical reactions associated with those cells. Active nerve cells require much more energy than inactive nerve cells. The body

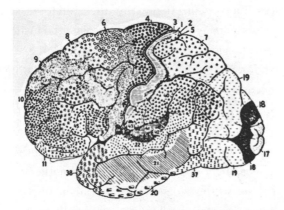

*Fig. 16.* A map of functional regions of the human cerebral cortex. Areas 17–19 are associated with vision, while 22, 41, and 42 are related to hearing. (Brodmann, *Vergleichende Lokalisation hehre der Grosshirnrinde in ihren Prinzipien dargestellt auf Grund des Zellenbaues* [1909])

responds to these increased energy demands by boosting the flow of oxygen-rich blood to the area of the brain experiencing an increase in nerve-cell activity.

When blood is oxygenated, it has different magnetic properties than when it is not. The slight differences in the magnetic behavior of blood with and without oxygen provide the signals that are measured by fMRI. An fMRI investigation thus shows us where activity is occurring in the brain and where it isn't occurring. It doesn't really tell us much about the nature of that activity. If aliens from outer space were spying on us, they might gather information about where important things were happening in a city by monitoring the flow of automobiles throughout the day. Cars flow from the suburbs into town during the business day and then to the restaurants and bars in the evening before finally returning to the suburbs, in each case indicating where important things might be happening, without giving any clues as to what those things might be. And so it is with the flow of oxygenated blood, as measured by fMRI.

In an fMRI test in which patients are shown images of human faces, an area of the brain now called the "fusiform face area" is often activated very strongly, much more so than when patients are looking at images of things other than faces. But specialization within the brain for tasks such as face recognition occurs at an even finer level, although the fusiform face area is a relatively small region within the brain. Specialization extends all the way down to the level of individual nerve cells. Studies dating back to the 1960s have shown that nerve cells can respond quite differently to various visual stimuli, such as a face versus some other familiar object. That specialization could occur at such a fine level in the brain was not well accepted, at least in the beginning. The MIT neuroscientist Jerome Lettvin, somewhat mockingly, coined the term "grandmother cells" to refer to nerve cells that are so specialized that they might respond only to the image of one's grandmother. If cells in the brain really are that specialized, Lettvin theorized, destroying a very small group of them might remove all memory of one's grandmother. You might still remember the smell of her apple pie, the look of her gingham dress, or the sound of her laughter, but she herself would be gone forever from your memory. That gingham dress would just be an empty shell, floating around the house, ghostlike. The term "grandmother cell" has entered the scientific lexicon and has even been formally reviewed by Charles Gross, another MIT neuroscientist, whose early work on specialization in nerve cells led, at least in part, to Lettvin's parable of the grandmother cell.

## Edges

Where things begin and where they end, their edges or boundaries, are important concepts in perception. When you look at a visually complex scene, such as a cluttered desk or a drawer full of kitchen implements, how do you make sense of the information your eyes are processing? Perceptually, it's crucial that you identify the edges of the individual items. Right now, I'm looking at a coffee cup on my desk. It's crammed full of pens and pencils along with other things such as a ruler, a flashlight, and a pair of scissors. Where does one item stop and the next one begin? Even if I didn't know what any of these items were, if I'd never seen any of them before, I would still be able to perceive them as individual items. My visual perception instantly finds the edges of each item and identifies them as such. The edges of an item are a property or feature that we perceptually identify, along with other things such as the item's color and surface finish. Visual edges are so much a part of our perceptual reality that it is hard to imagine their absence.

But edges are not limited to vision. They are hugely important in hearing as well. When we listen to a language that we understand, we perceive the beginning and the end of each word. Hereisawrittenexampleofwhatitwould-belikeifwecouldnotvisuallyperceivewhereonewordstopsandanotherstarts. So we put spaces between words in written language. Our brains learn to perceive those same spaces aurally. Even when we don't understand an individual word, we recognize its beginning and its end. The other day, my wife the attorney used the word "champerty" (attorneys love to throw around legal jargon).[5] I had no idea what that word meant. But I could find its edges. I could pluck it out from among the all the other words in the sentence and ask her what it meant.

Part of learning a new language is being able to perceive the edges of the words you are listening to. I have a gifted friend who speaks a half dozen languages fluently. Several years ago, he was preparing to spend a few weeks in Romania, where they speak a language he was unfamiliar with. He managed to find an online source for spoken Romanian, and he spent hours listening to it. When I asked him why, he said he was training himself to hear the beginnings and endings of the words. Once he can hear the edges of the words, he said, his oral comprehension of a new language starts to improve very rapidly. I think he's onto something. When I listen to a language I don't understand at all, such as Russian, it sounds like mush. All the words run together, and I can't tell where one stops and the next begins. I don't speak Spanish either, but as an American, I've listened to it often enough that I can often hear the edges of the words.

The neuroanatomist Jill Bolte Taylor describes her experiences with a massive stroke in her fine book *My Stroke of Insight*. In the days following her stroke, among many other problems, she experienced difficulty with edges of all sorts. For example, visual edges had disappeared for her. She had difficulty distinguishing the form of a person standing next to her bed, until that person moved. When a large conglomeration of pixels without any identifiable edges moved, she perceived the motion and concluded that that particular collection of pixels was a person. Her descriptions of her problems with the edges of spoken words remind me a little of learning a new language. She had to learn to find the edges of spoken English words all over again. Taylor describes another kind of perceptual edge as well—the edges of her own body in space. Part of the job of our proprioceptive

[5]*Champerty* refers to an agreement between a plaintiff and his attorney, in which the attorney agrees to finance a lawsuit in return for a percentage of any money won by the plaintiff.

and tactile senses is to help us keep track of where we end and the rest of the world begins. Taylor's stroke took away her ability to perceive those edges. Without them, she experienced a mystical feeling of oneness with space that most of us could scarcely imagine.

## Bottom-Up and Top-Down Processing

Imagine you've been invited to a fancy cocktail party thrown by your employer. It's important to look good, so you spend a little extra time selecting your outfit, taking care that the colors match and that nothing is stained or torn or dirty. Grooming is important, too—here the guys have it easier—and you make sure you look and smell just the way you want. Finally you arrive at the party. It's quite an affair. There must be several hundred people in that ballroom. The audio stimuli alone are nearly overwhelming. There's a jazz band setting the mood from a stage at one end of the room. On the floor, scores of conversations blend together, punctuated by peals of laughter and shouted greetings. Before you know it, your boss is tapping you on the shoulder and introducing you to her boss, who's in town from corporate headquarters. Time to make a good impression. Shutting out the music and all those other conversations, you concentrate on his every word. It's tough. The big boss didn't grow up here, and he has a wicked accent. He's had a few drinks, too, and that doesn't help. You hang in there, though, being sure to laugh at all his jokes and to answer all his questions quickly and politely. Finally, he moves on to meet someone else. You head for the bar yourself, secretly glad these affairs don't come around more often.

Even with all the computing power in the human brain, it seems like a miracle that someone can go to a party like that and make any sense out of anything at all. Stimuli impinge upon sensory receptors, creating electrical signals that move on to the brain. But there's no way we can consciously deal with all those stimuli and their resulting electrical signals. A lot of filtering and processing has to take place if we're to understand anything, much less a drunken boss with a foreign accent speaking to you at a loud party in a crowded room. It's useful to think of all the filtering and processing as occurring in two directions: from the bottom up and from the top down.

From the bottom up, unfiltered signals from the sensory receptors are decomposed into individual packets or streams of information. Signals from the hair cells in the organ of Corti, for example, are broken down into packets of information about loudness, pitch, duration, and location in space, among other things. That these data are separable in the brain shouldn't be surprising, since we can make the very same separations using human-made instruments.

From the top down, another whole set of processes is occurring. If the bottom-up processing consists of extracting features such as pitch, loudness, and location from sound stimuli (in the case of hearing), the top-down processes are more integrative.

Imagine that this book had been written in a language you don't understand. What would your brain perceive, as you look at these words on the page? Let's assume the words were written in a language that, like English, uses the Latin alphabet, such as German or Italian. Your eyes focus the light reflecting off the page onto your retinas, which then fire off electrical signals to the brain. There, bottom-up or low-level processing would differentiate the colors (black from white), the number and sizes of the black objects (the letters), and so on. Given that those black objects are letters in the familiar Latin alphabet, they would be recognizable as individual letters. But some highly sophisticated top-down processing is required to turn a collection of familiar black letters printed on a white page into something we can understand—something we call language. Top-down processing applies meanings to groups of letters or words. It integrates those meanings with the meanings of the words that came just before, and it anticipates what will come next. In English, if our processor perceives a noun, it will anticipate a verb. Things roll along that way, so that the printed page communicates with the human reader. This sort of top-down processing usually takes years to develop. It takes that long for your brain to organize itself such that it can handle this task.

Reading is thus a great example of how bottom-up and top-down processes work together in perception. We can read language on a page, but we can also hear it. Similar bottom-up and top-down processes occur when we listen to someone speak. Listening to language requires filtering out all the other auditory information that's arriving at the very same time. All those other conversations, and the music from the jazz band, for example, at that cocktail party. It also requires filling in missing information such as a swallowed or mispronounced syllable, or a word drowned out by a loud noise.

There are four elements involved in learning a new language: reading, writing, listening, and speaking. Each has its particular challenges. For my money, listening is the most difficult, although some of my friends disagree. But ask anyone struggling to master a foreign language, and they'll tell you. As difficult as listening in general can be, listening in a crowd is even worse. Deceptively, maddeningly worse. I learned this the hard way, during my first stint as a visiting professor in France many years ago. I'd been working hard to improve my rusty high school French for more than a year, and I was feeling naively confident about my language abilities when I stepped off the plane in Paris. Things got off to a good start when first I arrived at the little satellite campus of the University of Picardie in the town of Saint-Quentin in northeastern France. I met my department chairman, and we spent a good hour alone in his office getting to know one another. Not a word of English escaped our lips. The boss then introduced me to my new colleagues, and a little later I went to lunch with one of them. Another hour-long conversation all in French. The second day was similar to the first, and by its end, I was beginning to feel that it was all too easy.

Then came the third day. Lunch is the biggest meal of the day in France, and lunchtime is an important part of the French business day. It's not unusual to eat a big meal and remain at the table for an hour or more afterward, discussing business, sports, the weather, politics, and other such vital matters with one's colleagues. The French don't understand the typical five-minute on-the-go American business lunch, snatched from a fast food sack and inhaled in the car or at one's desk in total solitude. I grew to love the French way of having lunch, but I didn't love it that day. I soon realized that my one-on-one lunches on the first two days had been something of a test or tryout. A tryout for the real thing, when the number of people at the lunch table was not two but about a dozen. I couldn't believe what a difference this made. A conversation with two participants can move only as fast as the comprehension of the slowest person (in this case, me) allows. Add a third participant, and if two of them are native speakers of the language, the odd man out, the nonnative speaker, starts to get left behind. Add a fourth or fifth person—well let's just add twelve native speakers, divvied up into three or four different conversations. You begin to feel a bit like a large, dimwitted boulder, plopped down in the middle of a rushing river. Water—language—is flowing around you on all sides and over your head, carrying everyone along on its currents, except you. You can go from comprehending everything—every single word—in a one-on-one conversation to comprehending almost nothing in a large group like that with all those conversations rushing around you.

That was the realization I came to about halfway through that lunch. It hit me like a ton of bricks. My confidence, which had been soaring, vanished in a heartbeat, and I began to think I would *never* be able to communicate in a situation like that.

When you're learning a new language, you need every advantage you can get in order to understand and be understood. I do much better speaking French face to face than on the telephone, for example. On the phone, sound quality *always* suffers, and cell-phone connections, as we all know, can be really poor. Phone conversations also lack the little vital clues to meaning that are afforded by hand gestures and facial expressions. I've learned to watch people's faces carefully when they speak and to stay close to and in front of them when we converse. In other words, I've learned to use my eyes as much as possible to aid my ears and my brain. If I'm meeting someone at a restaurant, the acoustics of the place are at least as important as the menu. These things don't matter as much when I'm speaking my native English. My ability to process from both the top down and the bottom up in French isn't as good as it is in English. My brain isn't as good at things like finding the edges of words, filtering out extraneous noises, or filling in the missing pieces of conversations.

This last point, filling in missing pieces of the conversation, is important. The act of filling in missing pieces of sensory information in the perceptual stream, especially aural and visual information, is so important that it gets its own name: perceptual completion.

## Perceptual Completion

Perception involves drawing inferences from all the stimuli that we acquire through our sensations. Inferences are conclusions drawn from evidence. The evidence, in this case, is what comes into the brain through the senses. That evidence is first filtered and differentiated from the bottom up and then integrated from the top down. Bits and pieces of information are nearly always missing, though, and one of the most important tasks of top-down processing is filling in that information, or completing the perception. That this is so is easily proved through many examples of incidents when our perceptual software results in a demonstrably false conclusion.

The presence of the optic nerve on the retina creates a blind spot—a location in space from which an eye is not capable of processing the light it receives. This is one reason we have two eyes, since their respective blind spots do not coincide. The presence of the blind spot is quite easy to demonstrate, as shown in figure 17.

Lay the book flat on a table, open to the page showing figure 17. Now lean over the book and close or cover your left eye. With your right eye staring intently straight down at the large black dot on the left side of figure 17, slowly lower your head toward the book. When your head is

*Fig. 17.* A simple blind-spot demonstration exercise.

about eight inches or so from the page, the large plus sign will suddenly disappear. At that point, with your right eye directly above and focused on the black dot, the blind spot in that eye is geometrically lined up with the plus sign. You have demonstrated the presence of the blind spot in your right eye and, at the same time, observed a classic example of perceptual completion that leads to a false perception. If you keep lowering your head, a few inches lower the plus sign will suddenly reappear, for it will no longer be aligned with your blind spot. By moving your head up and down, a few inches in either direction, the plus sign will continue to appear and disappear. One could scarcely hope for a clearer demonstration of perceptual completion or of the respective roles of sensation and perception in our sensory world.

When your blind spot geometrically coincides with the location of the plus sign as you lower your head, the plus sign disappears. By perceptual completion your brain fills in the blind spot with what it infers should be there, in this case, blank white space, since that is what surrounds the plus sign on the page. Other demonstrations of the blind spot go further, using more complex geometric patterns to show more startling examples of visual perceptual completion. But perceptual completion in vision goes well beyond demonstrations of the blind spot. Top-down processing is constantly completing our perception of the visual landscape, filling in details in the periphery of the visual field (where visual acuity is lacking), compensating for poor lighting, and so on. It's such an integral, continuous part of visual perception that we don't even realize it's going on.

There are nonvisual examples of perceptual completion. The psychologist Richard Warren made an audio recording of someone reading this sentence: "The bill was passed by both houses of the legislature." He then recorded over a portion of the sentence with static, obliterating several words. The vast majority of those who listened to the altered recording reported that they heard both the words and the static, but most of them were unable to say where in the sentence the static had occurred. Their brains completed the sentence by filling in what they perceived should go in the space filled by static. Because this sentence is rather mundane and unsurprising, they could do so without error; it would be interesting to see the results of this study for native versus nonnative English speakers. In any event, living as we do in the cell-phone age, most of us are familiar with the perils of an overreliance on auditory perceptual completion: "Hi Mom. I'm at (static). I'll be home by (static)."

Most people in Warren's study reported that they heard the entire sentence about the bill being passed and somewhere along the line a burst of static. Having been differentiated from the bottom up, the static and the spoken words formed separate perceptual streams in these subjects' brains. The people who heard the missing words, supplied from the top down, were clearly imagining things, just as the blind-spot demonstration causes you to imagine things that aren't there. And yet this process is widespread and utterly human.

My wife is an attorney, and an excellent one, although I am hardly an unbiased witness. Most attorneys write a lot. Sometimes they have help editing their writing, and sometimes they don't. When she has to edit her own writing, my wife complains that she often does a lousy job and that errors inevitably slip through. In a recent brief, she had intended to write, "One of the plaintiff's strongest claims . . ." Instead, she wrote, "One of the plaintiff's strangest claims . . ." She told me she had read over that text, first on the computer screen, then in hard copy, perhaps a half dozen times without detecting this error. And since it's not a typographical or grammatical error, her word processor did not highlight it for her. Fortunately, the error was discovered by a colleague before the judge had a chance to see it.

"Why do I do that?" she wanted to know. "Why can't I see mistakes like that?" I tried to explain about perceptual completion, but I'm not sure I got very far. Perhaps this is one of those cases where a logical, analytical approach isn't called for. So, I just told her that we all make mistakes, that I love her very much, and that she is the best lawyer there ever was.

Editing your own work is difficult, as any writer will tell you. When other people edit your writing, they are much more likely to catch errors such as "strongest" versus "strangest." I suspect this is at least in part because, not knowing what you *intended* to write, they are more likely to see what you *actually* wrote. The urge to perceptually complete when reading one's own writing is very strong, not to mention strange.

# Sensory Deprivation

Our brains routinely fill in missing details in the perceptual stream. This is a complex process that develops over decades. The way I perceive visual information, as a middle-aged man, is vastly different from how I perceived it as a small child. I'm referring to the way my brain has developed over the years to process visual data—in other words, neural plasticity: my visual software, not my eyes or visual hardware. Perception is never entirely in the present. For centuries, philosophers have been fascinated by this fact, although, for the most part, they have had to be content to merely contemplate it, for it is a difficult thing to study empirically.

Difficult, but not impossible. Consider those who have been deprived of one sense or another for most of their lives and then have that sense restored.

There are relatively few cases of persons who are born blind or become blind at a very young age and then gain the ability to see many years later. The neuropsychologist Richard L. Gregory reported on the case of S.B., a man who was blind from the age of ten months until age fifty-two. S.B.'s nonfunctioning corneas were replaced in 1966 by a functioning pair from an organ donor. Before regaining his vision, S.B. had led an active life as a blind man. He enjoyed gardening and going for walks, and even bicycling, with a sighted cyclist to guide him. He had long dreamed of rejoining the sighted world, of which he had no conscious memories, although physicians had deemed his case nearly hopeless. When his cornea transplant was finally performed, it was successful. But S.B.'s story ends in tragedy.

A few days after the surgery, he could see reasonably well. He could navigate the corridors of the hospital by sight, he quickly learned to tell time from a wall clock, and he enjoyed watching the street scene from his hospital room window. However, not everything was normal. His perception of distance was distorted. His hospital window was at least thirty feet above ground, yet he believed that if he lowered himself out the window by his hands, his feet would touch the ground.

Within a few days of leaving the hospital, S.B. began to experience periods of depression, which became more and more frequent and severe. Eventually, he gave up all attempts at an active life, and he was dead three years after the surgery.

While this is an extreme case, depression in individuals who have gained vision or regained it after years of blindness is common. Little by little, S.B. began to find the world revealed to him by his vision wanting. He liked bright colors, but fading or flaked paint and other blemishes disturbed him, as did the failing light at the end of the day. Eventually, not unlike others in similar documented cases, S.B. reverted to living much as he had when he was blind. Often, he would not even bother to turn on the lights at his home in the evening.

Several anecdotes provide insight into what sighted life was like for S.B. Not long after his surgery, he was shown a lathe, a machine whose purpose he understood and one in which he had expressed great interest. At first, he was incapable of visually deducing anything about how the machine might operate. He could only say, tentatively, that one particular component looked like a handle. He then closed his eyes and ran his hands rapidly and expertly over the entire machine for more than a minute. Reopening his eyes, he reported, "Now that I've felt it I can see."

When S.B. was blind, he confidently navigated his neighborhood on foot with his white cane, and he was not intimidated by vehicular traffic. At crosswalks, he would stand at the curb and hold his stick out horizontally into the street. When his ears told him the traffic had stopped, he would boldly plunge into the street and cross to the other side. After the surgery, however, it was a different story. He was extremely distrustful of his vision and became terrified of crossing a street. Even with an aide at each shoulder, he could scarcely bring himself to take the first step.

Could it be that, lacking the decades of visual "software development" that most of us have gained, S.B. was simply overwhelmed by the flood of information from his eyes—his suddenly functioning visual sensory receptors? What a powerful phenomenon this must be, leading as it did to severe depression.

## Molyneux's Problem

The sad case of S.B. brings to mind Molyneux's problem, which has intrigued philosophers since it was first posed in 1688. The Irish scientist and politician William Molyneux (1656–1698) described the problem in a letter to the esteemed English philosopher John Locke (1632–1704). It goes like this: A person, blind from birth, learns to distinguish many objects through the sense of touch. For example, when he is given a small wooden cube and a wooden sphere of the same size to hold in his hand, he can tell which is which. Now, imagine that this blind person is suddenly made to see and that he is then shown one of these objects. Would he know, based on sight alone, which one it was, the cube or the sphere?

Locke, no mean thinker, declared that Molyneux's problem was a dandy. Locke published the problem, and subsequently a veritable who's who of intellectual giants took swipes at it, including Voltaire, Gottfried Leibniz, Denis Diderot, George Berkeley, Hermann von Helmholtz, and William James. Some, like Locke and Molyneux himself, believed the answer was no: the formerly blind person would be unable to distinguish the sphere from the cube based on sight alone. Others, including Leibniz, thought the answer was yes.

There was no consensus, in part because the problem as published by Locke can be interpreted several ways. Having written thousands of exam questions in my day, I am painfully familiar with the hazards of misinterpretation. Molyneux and Locke make no mention of whether the subject should be given time to reflect on his answer or be made to answer right away. They also differ on whether the subject should be told that he will be shown either a sphere or a cube and asked to tell which it is, or whether he should simply be shown "an object" and asked to name it.

Things got even murkier when the first empirical evidence came along, in 1728. The English surgeon William Cheselden (1688–1752) removed extensive cataracts from the eyes of a thirteen-year-old blind boy, restoring his vision. Cheselden reported that, after the surgery, he had posed Molyneux's problem to the boy, who was unable to distinguish the sphere from the cube.

The philosophers had even more fun with Cheselden than they'd had earlier with Molyneux and Locke. They pointed out, among other things, that the boy may not have had enough time to convalesce after surgery, that he may have had *too much* time, and that cataract sufferers such as this young boy are rarely profoundly blind, so this case may not have provided a fair test of Molyneux's problem. Some suggested a rigid set of rules to follow the next time such a case presented itself, including a prescribed period of convalescence in a pitch-black room, before the problem was posed. The welfare of the patient seems to have been lost to some of these intellectuals, in their zeal to solve a vexing little brain-teaser.

Subsequent patients exposed to Molyneux's problem (including the case of S.B. from the 1960s) haven't cleared things up much, and there remains no consensus on the answer to the problem. What does seem clear is that much of what we call vision is learned, or acquired through experience. As I write these words, I am within days of my fifty-second birthday. S.B. was fifty-two when his vision was restored. What would my life be like, what sort of person would I have become if, like S.B., I had lived nearly fifty-two years with nonfunctioning eyes? I cannot even begin to imagine it.

## Neural Plasticity

Notwithstanding an extreme case like S.B., it is clear that the human brain has a remarkable ability to adapt to changes in sensory inputs. The brain, over a period of weeks, months, or years, can actually restructure itself; for example, it creates new connections among nerve cells through a process called neural plasticity.

The intricate details of the central nervous system are far too complicated to be determined entirely by genetics. The brain is not hard-wired,
as were the primitive computers of the 1950s. Genetic changes can't take
place any faster than from one generation to the next, and in practice they
occur over many generations. That a living brain has the ability to rewire
itself as it learns new things is what gives us the ability, within a single
lifetime, to master a new technology, speak a new language, or compensate for changes in one of our senses—or even the loss of a sense.

The brain can't rewire itself to do something that the sensory hardware isn't capable of. We can't learn how to see ultraviolet images, the way
a honeybee can, or to hear high-pitched sounds like a dog. To gain these
abilities would require genetic—evolutionary—changes. But humankind
has managed to learn how a bee's eyes work, about their ability to see in
the ultraviolet range. And we've created instruments that capture ultraviolet images and allow us to see what a bee sees. In such ways the mag-
nificence of the human brain has allowed us go far beyond the limitations
of our sensory hardware.

Likewise, our brain can't relearn how to hear if the hair cells inside the
organ of Corti no longer function. Unless, that is, a person with such a loss
receives a cochlear implant, a miraculous device that uses human-made
electronics to directly stimulate the auditory nerves.

Interest in the direct electrical stimulation of hearing dates back at
least to Alessandro Volta (1745–1827), the inventor of the battery. Volta's
approach was direct, if somewhat inelegant. With electrodes stuck in his
ears, he subjected himself to jolts of direct current. He reported hearing
noises that sounded like boiling soup. The first recorded instance of direct
electrical stimulation of auditory nerve cells occurred in the 1950s, when
the French physicians André Djourno and Charles Eyriès implanted wires
in the inner ears of patients. Upon electrical stimulation of those wires, the
patients reported hearing distinct sounds like crickets chirping or rou-
lette wheels turning. Others repeated and confirmed these results, and the
race was on to create what we now call the cochlear implant.

The cochlear implant depends on a lot of sophisticated technology for its ability to restore hearing to some profoundly deaf individuals. One factor most crucial to the success of the cochlear implant, however, is not human-made. That factor is neural plasticity, the ability of the brain to rewire itself over time.

When Michael Chorost's cochlear implant was first activated, he could make very little sense of its electronic signals. The noises he perceived were a bizarre cacophony of electronic buzzes, hisses, and squawks. He could make no sense of spoken language. What had once been the familiar sounds of his life were now maddeningly unintelligible. Slowly, over weeks and months, he began to be able to comprehend the auditory information the new system provided him. "The software [of the cochlear implant system] had not changed," he notes in *Rebuilt.* "The world, presumably, had not changed. What had to have changed was my brain."

The interface between a cochlear implant and the central nervous system is called the electrode array (figure 18). It is a very flexible rod about a millimeter in diameter and an inch or so long. Spaced along the rod are a series of about twenty-four platinum electrodes. Each electrode is responsible for transmitting to the brain electrical signals corresponding to sounds in a different frequency range. Those electrical signals begin as sound waves collected by a microphone attached to the patient's head, as described earlier.

As recently as the 1970s, some experts believed that direct electrical stimulation of nerve cells could not provide meaningful information to the brain. But that is exactly what a cochlear implant does. The trick is to electrically stimulate the nerve cells in ways that closely approximate the actions of the cochlea itself in individuals with normal hearing. "Closely" won't be close enough, at least at the beginning, as Chorost's reactions show. But neural plasticity can save the day.

The brain changes, throughout life, by several mechanisms: strengthening connections, adding new connections or removing old ones, and generating new cells. Taken together, these mechanisms constitute neural

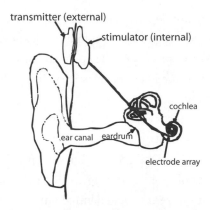

*Fig. 18.* A cross section of the ear showing the electrode array of a cochlear implant coiled up in position inside the cochlea.

plasticity. The details of how all this happens are the subject of much research. It is well known that the brain changes easily and rapidly during certain periods of development, particularly when we are young. The brains of infants can create up to 1.8 million new nerve-cell connections per second. Learning a new language is far easier for the very young than for others, as study after study has shown. The ability to discriminate the speech sounds of a new language goes through a "critical period" from about six months of age to one year. After that, it becomes much more difficult. Other aspects of language learning do not begin to diminish until much later, it appears. But learning a new language fluently is far more difficult, for most of us, after the middle teenage years.

Critical periods for all sorts of developmental tasks have been identified. During these times, the ability of nerve cells to make connections related to a given skill or ability is at its peak, and the wiring patterns established during that period are relatively permanent. Many studies have been done on critical periods, and lots of those studies relate to vision. If, for example, one eye of an adult cat is covered for several months and then uncovered, there are no long-term effects on the cat's vision. However, if one eye of a kitten is covered for the first two to three months of life, the cat is essentially rendered blind in that eye. The connections in the primary visual cortex that normally would be shared by both eyes are taken over by the eye that was left uncovered. The eye that had been covered, even though it is normal, is ignored by the brain, and the effect is permanent. It is not for nothing that these are called "critical periods."

Plasticity is thus a two-edged sword. It is an enormously powerful tool that has evolved to allow us to adapt to a changing environment. But if the proper stimuli are not present during those critical periods, the result can be permanent disability.

Plasticity, however diminished it might be by critical periods, continues throughout our lives. If ways could be found to reactivate the high levels of plasticity that occur during critical periods, the discovery could lead to new ways of treating a range of neurological disorders, as well as perhaps aiding people who have lost a sense.

## Synesthesia

When Isaac Newton performed his famous prism experiments in 1666, splitting a beam of white light into a spectrum of colors, he concluded that there were seven colors in the beam: red, orange, yellow, green, blue, indigo, and violet.[6] As we have seen, we can perceive many more than just seven colors in that spectrum. Nonetheless, Newton selected seven, and in doing so he drew an analogy to the seven notes of the musical scale. Was this merely a convenient coincidence for Newton to latch onto, or did he perceive a more tangible connection between music and color? Was Newton a synesthete?

*Synesthesia* can be defined as perceiving sensations from one sense, for example, hearing, with another sense, such as vision. A synesthete might perceive the color blue when he hears a piece of music in the key of D major, whereas G major might evoke yellow. The word *synesthesia* shares

---

[6]These days, as noted in part 1, indigo is generally not included in the visible spectrum.

a root with *anesthesia*. The latter means "no sensation," and the former means "joined sensation."

Connections between music and color are certainly not limited to synesthesia. Music can be "dark" or "light," and "the blues" is a distinct form of musical expression. But for the true synesthete, the association goes well beyond the metaphoric. For the musical synesthete, a piece of music in a certain key is not *like* red, it *is* red. Explaining synesthesia to someone who isn't a synesthete, that is, to most of us, is a little like explaining vision to someone who has been blind since birth. Where do you begin?

One place to begin is by noting that color, even in the normal sense of things we see, is a figment of one's imagination, which is to say that it is a perception. The red of a stop sign is something that exists only inside our brains. The neural connections in our brain that cause us to perceive "red" when we see a stop sign are considered "normal." Those same connections must somehow get made when a musical synesthete perceives music in a certain key as red.

There are many different kinds of synesthesia, perhaps as many as sixty. In researching the subject, I was startled to learn that the way I perceive time can be considered a form of synesthesia. For as long as I can remember, the yearly calendar has existed in my mind like a huge ellipse suspended in the darkness of space. The details of this picture never change, and to me they are so real that I feel I could reach out and touch them. The months of the year are arranged uniformly around the ellipse. The closest month to the observer (me) is June, which happens to be when I was born. Moving slightly up and away, to the left, you get to July and August, then the ellipse turns down slightly toward September and October, and more dramatically to November and December. The New Year begins near the low point on the ellipse, and by February, things are turning up again. If the ellipse were a clock face, time would thus be moving counterclockwise. I cannot remember ever perceiving the calendar in any other way, and I recall thinking it odd the first time I realized that not everyone perceives time this way. In similar fashion, historical time and the weekly calendar appear in three-dimensional space in my mind, except that the geometric form is different. This type of synesthesia, one of the milder and more common ones, I suspect, is called spatial-sequence or number-form synesthesia.

Estimates of the number of synesthetes vary wildly from as many as one person in twenty to as few as one in two thousand. Several factors combine to make it difficult to estimate the frequency of synesthesia as compared to a condition like color blindness. First, there is no general agreement on exactly what constitutes synesthesia. Second, many people, and I would have included myself until recently, are unaware that their synesthetic perceptions are in any way unusual.

Investigations into synesthesia were popular in the 1800s. Such studies went out of fashion for much of the twentieth century but of late have been experiencing a renaissance. This has coincided with the rapid development of brain research tools such as fMRI, which allow scientists to observe which parts of the brain are stimulated during various mental activities.

The redoubtable Sir Francis Galton investigated a variety of synesthetic phenomena in his *Inquiries into Human Faculty and Its Development* (1883). He recorded his observations of individuals who saw letters of the alphabet or numerals in different colors. A given letter or numeral always had the same color, whether the person observed it on a page or imagined it. This is sometimes called grapheme-color synesthesia. The Nobel Prize–winning physicist Richard Feynman writes about his own experience with this in his 1988 memoir, *What Do You Care What Other People Think?* Patricia Duffy describes learning how to write the alphabet as a young girl. "I realized that to make an R all I had to do was first write a P and then draw a line down from its loop. And I was so surprised that I could turn a yellow letter into an orange letter just by adding a line."

Galton and others concluded that these synesthetic perceptions were real, not associated with any sort of learned behavior, and virtually impossible to influence through the conscious mind. In other words, if the letter A is red to a grapheme-color synesthete, the person cannot make it blue or any other color, any more than you or I could, through willpower alone, make sugar taste like salt.

There are, however, synesthetic phenomena involving taste, as described by the neurologist Richard Cytowic. When a cook apologized to him for not having put enough "points" on the chicken he had prepared, Cytowic realized that this cook's sense of taste was different: it was somehow synesthetically connected to his sense of touch. Cytowic became an expert on synesthesia as a result of the experiences he chronicles in *The Man Who Tasted Shapes*.

## Strokes and Perception

A stroke is a loss of brain function that results from some sort of disturbance to the brain's blood supply. Stroke is the number three cause of death in the United States and the leading cause of adult disability. Stroke survivors face long, difficult rehabilitations and sometimes report feeling like a different person after their stroke—whether or not they suffer any permanent disabilities of the traditional sort, such as paralysis or vision difficulties. Different, perhaps, because their brains will never be the same as they were before they were injured by stroke. In *My Stroke of Insight*, Taylor notes that some of her doctors told her that if she hadn't recovered a certain ability by six months after her stroke, she would never recover it. Taylor concluded that this was absurd, since she noted improvements in various aspects of brain performance for fully eight years after the stroke, by which time she had recovered nearly every ability she had had before the stroke.

Depending on which region of the brain is afflicted, strokes can destroy or degrade just about anything the brain does, including various aspects of perception. Taylor's perceptual problems with various kinds of edges were discussed earlier. Some of her other perceptual difficulties are so unusual that they can't help but change the way one thinks about how the brain must work. For example, not long after her stroke, as Taylor struggled to recover, she was given a simple twelve-piece child's puzzle to work. It was almost more than she could handle, so her mother helped her, noting at one point that she could "use color as a clue." This was a revelation. As soon as Taylor heard her mother's words, she could suddenly see color again. "It still blows my mind (so to speak) that I could not see color until I was told that color was a tool I could use. Who would have guessed that my left hemisphere needed to be told about color in order for it to register?" Taylor describes similar perceptual revelations in which a certain ability had to be waked up or otherwise revealed to her conscious mind before she could utilize it.

## Image Reconstruction

Jill Taylor's descriptions of how she learned to piece back together her perceptual world after her stroke remind me of the scientific field of image reconstruction.

In medicine, there exists a veritable alphabet soup of powerful imaging technologies, ranging from MRI (magnetic resonance imaging) to CT (computerized tomography) to PET (positron emission tomography). All of these technologies involve creating a visual image from measurements of various types of electromagnetic waves or radiation. While the details of the different technologies vary wildly, what doesn't vary is the need for image reconstruction. What these instruments are measuring isn't "visual" in the normal sense. For example, MRI is the mapping, into a color image, of varying levels of emissions in the radio-frequency range by subjecting different kinds of tissue to an intense magnetic field and electromagnetic pulses. The color image is simply a much more convenient way to perceive the torrents of information flowing out of an MRI machine than the raw numbers would be.

The measurement by the medical instrument is always both incomplete and inaccurate. Some portion of the body part being imaged is not fully represented (incompleteness), and the signals from the instrument will always contain a certain amount of "noise" (inaccuracy).

If this sounds a lot like our earlier descriptions of how perception works in humans, it should. These factors—incompleteness and inaccuracy—are always present in our natural sensations as well. Easily demonstrated phenomena such as the eye's blind spot illustrate this fact, but such special cases tend to obscure how utterly normal it is for our brains to have to deal with incomplete and inaccurate sensory information. Image reconstruction is needed both by the human brain attached to a pair of human eyes and by a computer attached to an MRI machine.

# Multitasking

Publilius Syrus, a Roman slave in the first century B.C., observed, "To do two things at once is to do neither." Publilius sounds like my kind of guy. I'm a lousy multitasker, too. When I'm driving a car in a strange city and realize that I am lost (a not-infrequent occurrence, I'm afraid), I am suddenly compelled to turn down the volume on the car radio. I suppose I do this because I need to concentrate, to get my bearings, so that I can find my way. I want less extraneous sensory input, which is what the radio represents. I want to concentrate on what my eyes tell me, to bring in as much information visually as I possibly can (road signs, landmarks, the location of the sun, etc.), and to concentrate my perceptual abilities on these stimuli. Turning the radio down is not so much a conscious decision as a reflex. As long as I know where I am going, I can relax while I drive and listen to the radio, but as soon as I realize I'm lost, I have to turn it down.

Electrical engineers spend a lot of effort making sure that none of the various wired and wireless signals passing through modern machines (cars, computers, airplanes) are interfering with any of the other signals. I think something like that must go on with our perceptions. The first time I installed a car stereo, many years ago, it sounded just great—until you started the car's engine. Because the stereo wasn't properly shielded from the engine's ignition system, you could hear the engine, or more precisely the ignition system, through the stereo speakers. This noise pretty much drowned out the music coming through the stereo. I needed help to fix this problem, and perhaps that explains why I'm now a professor of mechanical engineering, and not electrical. We are fortunate that modern electromechanical devices such as cars and airplanes are designed such that—at least until an amateur interferes—the myriad electronic signals they generate don't interfere with one another.

When I try to do something on my computer while talking on the phone, the quintessential modern multitasking scenario, I end up doing neither task very well. I'm not ashamed to admit, likewise, that I'm not very good at talking on my cell phone while driving. Driving while using the phone is illegal in lots of places, and it would be fine with me if it were illegal everywhere. Study after study, including those using some very sophisticated driving simulators, has shown that driving while using a cell phone is about as dangerous as driving while drunk. According to research done by Ford, a cell-phone conversation impairs a driver in a manner that is statistically similar to having consumed three alcoholic beverages. And text messaging while driving is even worse.

The debate over cell phones and driving is reminiscent of a similar squabble in the 1930s, when car radios began to become popular. Back then, there were people who thought radios shouldn't be allowed in cars because they would distract drivers and thus cause accidents. We all know who won that particular battle. Cell-phone talking and texting are different from simply listening to the radio, since they involve the driver's hands and eyes, not just his ears. Will the outcome be different this time? Only time will tell whether driving while "intexticated" will be deemed a violation of the traffic laws or simply a matter of personal freedom. The freedom to be an idiot, I suppose.

*Multitasking* is a term that has come into general use from its genesis in the computer industry. According to an online etymology dictionary, the term was coined in 1966. One definition of *multitasking* in computers is the ability of an operating system to run more than one software program at the same time. Before the late 1980s, personal computers could run only one program at a time. If you wanted to use a computer spreadsheet, you had to quit the word processor you were running and then open the spreadsheet separately. Multitasking operating systems allowed for multiple programs to be open simultaneously, a development that made the personal computer much more convenient to use, and today we all take this feature for granted.

But computer scientists tell us that a single central processing unit can really do only one thing at a time. A computer's ability to multitask is thus somewhat of an illusion, because what the computer really does is to interrupt one program to work on another, drop that to deal with a third, and so on. Computers work so fast that they appear to be doing things simultaneously, when in reality they are doing them sequentially.

Research on the ability of humans to perform multiple tasks in parallel dates back several decades. Results vary, but one general conclusion is that most people aren't nearly as good at multitasking as they think they are.

# Chapter 9 • Perception and Culture

Stimulus and sensation are black and white—color vision notwithstanding. Perception is in shades of gray. Stimulus and sensation are the stuff of science: observable, factual, and unchanging. Perception is the stuff of science too, but it's the science of the mind. The science of Sigmund Freud, not of Isaac Newton. Earlier, we met the durian, a strange fruit to Western eyes and noses. The molecules given off by the durian are the same in Vietnam as they are in Indiana. So why is the durian the King of Fruits in Southeast Asia and a disgusting abomination in many other places? It's all a matter of perception and culture.

## The Culture of Eating

We like to eat in America, and we know what we like in a plate of food. It is my opinion, however, that in America we don't really have an "eating culture." It's hard to realize this until you've spent time in another culture, someplace where the eating experience is culturally more important. Several years ago I was visiting friends in Paris, and we decided to take a weekend trip to Brittany. As we checked into our hotel in the lovely town of Saint-Malo, my Parisian friends asked the desk clerk about the local restaurants. What followed amazed me. For a good ten minutes, we discussed the local culinary scene with the teenage desk clerk. Even though we were old enough to be her parents, this young girl, in spite of her youth and her collection of body piercings, tattoos, and a spray of multicolored, spiky hair, seemed totally at ease discussing the pros and cons of at least a half dozen restaurants within walking distance of our out-of-the-way hotel. Such-and-such restaurant was likely to have the best selection of fresh fish (Saint-Malo is on the English Channel), and the chef there loved to prepare it a certain way, which she described in detail. The chef at another place favored a different style, which she also described. All aspects of the dining experience were discussed: which restaurants had the best wine cellars, the most interesting cheeses, the most delectable pastries. Atmosphere was not neglected, nor was price.

At the end of this discussion (I was relieved there was no one waiting in line behind us), my friends had developed a short list of three candidate restaurants for our evening meal. It was a lovely summer day, and on our subsequent sightseeing promenade, we passed by each of them. All were closed, which is common in France between the hours of lunch and dinner, but we were able to peer in the windows and otherwise size up the locations. Finally, a selection was made.

When we returned for dinner that evening, we weren't disappointed. The food was superb, the service impeccable, and the atmosphere warm and inviting. I ordered the "catch of the day," a fish whose French name I did not recognize. Out of simple curiosity, I asked my Parisian friends if they knew the English name of this fish. They did not. My inquiry was passed on to the waiter, who did not know the English word, either. Neither the other waiter in this small restaurant nor the young girl busing tables could help us. Soon, the chef himself was standing beside our table. He was similarly unable to provide a translation. The fish was delicious, and by that time I couldn't have cared less what its English name was. I was feeling more than a little embarrassed to have brought it up in the first place. Nonetheless, the quest for a translation continued. The chef whipped out his mobile phone and was calling friend after friend. The other patrons in the restaurant were being consulted as well. Finally, the chef struck gold. A fellow chef, in Paris no less, knew the answer: cod (*morue* is the French name). By any other name, it would have been just as tasty.

What is the point of this story? If you're thinking, "The point is, the French are food snobs," then I haven't done a very good job of telling my tale. What this story means to me is that this is just how things go in an eating culture. In such a culture, food—how it tastes and smells, how it looks, how it is prepared, how it is served, where it comes from, *everything* about it—is vitally important. If their American customer wanted to know the name of the fish on his plate in his native language, the staff of this restaurant was going to do their very best to find it out for him, because knowing what you are eating is important.

And what of the young clerk at the hotel—she of the detailed discourse on the local cuisine? I decided to do a little experiment. The next few times I stayed in an American hotel, I made it a point to ask the desk clerk for restaurant recommendations. In each case, I was simply and silently handed a photocopied map with a set of directions guiding me to "Restaurant Alley," where I was certain to find, lined up along either side of the road, Chili's, The Olive Garden, Friday's, P.F. Chang's, and an assortment of other chain restaurants whose names are repeated with mind- and palate-numbing sameness from one American city to the next. So omnipresent are these establishments that there was no reason for the desk clerk to wax eloquent on the epicurean delights that awaited me at any of them. I might just as well have asked for recommendations of where, in order to sample some of the local culture, I might go to fill my car with gasoline. "Sir, the Shell station down the street serves a lovely 91 octane this time of year. You really owe it to yourself to try it out." Filling our stomachs, like filling our gas tanks, is, for many of us, a pedestrian task. Convenience rules, with cost a close second. Too often, taste gets left out of the bargain.

Michael Pollan, the author of several important and insightful books about food, considers the culture of eating in America in his book *In Defense of Food*. He notes that our Puritan roots tend to "impede a sensual or aesthetic enjoyment of food." Another contributing factor could be the "sheer abundance of food in America." That good fortune may have rendered us somewhat indifferent to food, so that we prefer the eat-and-run lifestyle, as opposed to lingering at the table and truly enjoying the dining experience. The eating habits of various cultures, says Pollan, are among "the most powerful ways they have to express and preserve their cultural identity."

## In the Dark

At *Dans le Noir,* a Paris restaurant that also has locations in other cities, including London and New York, the waiters are blind. All the better to serve their clientele, who, for the duration of their meal "In the Dark," are blind as well. All meals at *Dans le Noir* are served and consumed in rooms of profound, absolute darkness. This is a gimmick, but one with several goals behind it. Chief among them is to foster relations between the worlds of the seeing and the nonseeing, to give the sighted some idea what it is like to be totally blind, if only for an hour or so. But beyond that, *Dans le Noir* sees its mission as helping its clients "completely reevaluate their notions of taste and smell," according to its Web site, danslenoir.com. In an atmosphere of profound darkness, the eye is powerless to influence the dining experience.

Anyone who doesn't believe that the eye can be more powerful than the tongue when it comes to taste has never lived with children. "That looks disgusting. There's no way I'm eating that." is a cry all too familiar to parents everywhere. And another meal, lovingly prepared, ends up in the garbage. At *Dans le Noir,* this can never happen. There, you can smell your meal, taste it, listen to it, or even feel it with your fingers (who would ever know?), but you cannot see it. There is no need for Red Dye No. 2 in anything, or for farm-grown salmon that has to be artificially colored orange. Your meal must win your favor through your nose and your tongue, not through your eyes.

The eyes play at least one other role when it comes to food, and that is in the evaluation of portion size. The eyes, as the saying goes, are sometimes bigger than the stomach.

## The Portion Wars

One of my favorite scenes in the Mel Brooks film *Blazing Saddles* comes near the very end of the picture, when, at the concession stand in a cinema, melted butter is dispensed into a garbage-can-sized container of popcorn through a nozzle designed to look like a gasoline pump. Movie houses love to sell us popcorn. It's more than love. They *have* to sell us popcorn, if they want to survive financially. Happily for them, the stuff is quite profitable. The former manager of a multiplex told me that the profit margin on movie-house popcorn is well over 90 percent. So why not sell it in ever-increasing, and expensive, portion sizes? Movie popcorn, at least in America, has been sold in embarrassingly large containers for decades now (*Blazing Saddles* was released in 1974). But popcorn was ahead of its time. Nowadays, just about everything has been supersized. Although discredited terms like *supersized* may have gone out of fashion, thanks in no small part to Morgan Spurlock's 2004 movie *Super Size Me,* the trend toward increasing portion sizes in restaurants continues. Wendy's dropped the size labels "Biggie" and "Great Biggie" in favor of the more pedestrian "medium" and "large" but did not correspondingly decrease the size of its portions. A rose by any other name would smell as sweet, and a gargantuan tub of fries is just as large no matter what you call it.

In 1996, a colleague of mine walked into a party carrying a soft drink, a plastic cup with a straw, which he'd just purchased at a convenience store. It was rather large, probably thirty-two ounces or more, but I thought nothing of it. It appeared quite ordinary to my American eyes. Attending the party was a young German exchange student who'd just arrived in the United States. He took one look at that bucket o' soda and his eyes opened wide. "I come back right now," he informed us in his rapidly improving English, as he bolted out the door to fetch his camera. "My friends in Germany never believe this unless I make a photo," he exclaimed, and my colleague, beverage in hand, congenially mugged for the camera.

It's easy to see why movie houses have oversized popcorn, why convenience stores oversize beverages, and why restaurants have followed suit with their meals. It's a lesson in economics coupled with perception. Let's use movie popcorn as an example. What does it cost the multiplex to provide you with that enormous tub of popcorn? A few pennies for the unpopped corn, another penny or so for the foul-smelling potion they call butter, and a fraction of a penny for salt. The container itself, at a nickel or more, probably costs at least as much as all the stuff that goes inside it.[1] Throw in another penny or so for the electricity needed to pop the corn. And don't forget the fresh-faced high school kid who takes your order and serves you the popcorn, the fancy stainless steel machine the stuff is popped in, and the person who (sometimes) cleans up the popcorn mess in the cinema after each show. Those are important costs, too.

Economists sometimes divide these costs into two categories: variable and fixed. Variable costs vary with the amount of product sold. Every time you serve a customer, that's another nickel for a container. The container is a variable cost. The kid who takes your order, however, is a fixed cost. He makes the same hourly wage no matter how many, or few, orders of popcorn he sells. And the popcorn machine has the same purchase price regardless of how much you end up using it.

In the food service industry, the fixed costs are often more important than the variable ones. Popcorn at the multiplex is a great example. The variable costs (corn plus butter plus salt plus electricity plus the container) might be only about a dime. So here's where sensation and perception enter the equation. For a few pennies more (a little more unpopped corn, butter, and salt, plus a bigger tub), I can provide the customer with some serious eye candy: a giant vat of popcorn so big you have to carry it with both hands. By going to the bigger size, I win two ways. First, I sell more product, since people can't resist the sight of that huge bucket. Second, since the portion is larger, I can jack up the price. If I charge four dollars for a one-gallon container of popcorn, then I might sell a two-gallon serving for six

---

[1] Far from being an anomaly, that's true as well for lots of packaged food, including most canned or bottled beverages, such as beer, soda, and water.

dollars. The difference, two dollars, is almost pure profit. This is the economics of portion size.

It's not so surprising that this has happened as that it took so long to happen. As more and more of the food service business has been taken over by chains (multiplex cinema chains, convenience store chains, fast-food chains, and restaurant chains), portion sizes have grown. The economic principle is relatively simple, and now it has been embraced throughout the industry.

Huge portions have gone from a clever way to increase profits to an industry necessity. Restaurants have come to believe that they are judged on their portion sizes just as surely as they are judged on the quality of their food or on atmosphere, service, or any other factor. It is only, ironically enough, in the very most expensive, exclusive dining establishments where small portions are tolerated or even expected. Anywhere else, you'd better dole out the servings with a shovel, lest your clientele abandon you.

Eventually the dining public began to notice, and in the light of the very real obesity epidemic in America, the Portion Wars really heated up, and a backlash developed against oversized servings. In June 2006, the U.S. Food and Drug Administration published portion-size recommendation for restaurants. On the face of it, this is ridiculous. Imagine if you went to the filling station to buy gasoline for your car and were informed that, due to recommendations from the Department of Energy, fuel was available only in one-gallon servings.

Many in the food service industry, and elsewhere, argue that the government has no business telling restaurants, or even officially recommending, how much or how little food they may serve their customers. They maintain that the logical, intelligent dining public is fully capable of controlling itself and that there are at least two excellent ways of dealing with a too-large portion: the shared plate and the to-go box. This is all true. Sharing a restaurant meal makes great sense, assuming you're not dining alone. Even though many restaurants add a charge for splitting a dish into two servings, it still makes sense both economically and calorically to split meals. Portion sizes have grown so large that even after splitting a single meal, my wife and I sometimes have enough leftovers for a to-go box.

Here's another trick (this one works best when you are dining out with friends and have, say, a table of four): Order two meals, and ask the waiter to serve them to you in to-go boxes while also providing each diner with an empty plate. The to-go boxes then are serving dishes, which can be passed around as everyone takes what he or she wants, family-style. Everyone gets to choose which dishes, and how much of each, to eat. And when you're done, the leftovers are already in the to-go box. On the downside, I will admit that those huge styrofoam clamshells can detract, if ever so slightly, from the atmosphere of a fine dining establishment, with its bone china, linen table cloths, and crystal wine glasses.

To-go boxes make economic and caloric, if not aesthetic, sense. At my house, however, they frequently spend several days cluttering up the fridge before departing, unopened and unmourned, with the garbage. Nonetheless, splitting meals and to-go boxes remain sound, logical strategies for winning the Portion War. Or are they? Study after study has shown that many people will eat much more if much more is available.

This is a complicated problem, driven by sensory, cultural, and instinctual factors. In our culture, as children, we're all taught to eat what's on our plates. Ironically, those lessons don't seem to take effect until we become adults. So larger portions translate into bulging waistlines. The instinct to eat more when more is available may be the most powerful force of all.

Snoopy, the beloved beagle creation of the late Charles Schulz, is sleeping on his doghouse one day, when Charlie Brown arrives at suppertime with not one bowl of dog food, but two. He informs Snoopy that he will be out of town the next day. The second supper dish, he says, is for tomorrow. Charlie Brown advises his canine friend not to be greedy and eat both bowls at once. Snoopy dutifully descends from his perch, eats one bowl of food, and then tries to go back to sleep. But he can't stop thinking about that second bowl of food. Finally he can stand it no longer, and, leaping from his doghouse, he devours the second portion. "I'd have hated myself if tomorrow never came," he explains as he resumes his nap, satiated at last.

Snoopy's problem here is ours as well. There are those who believe that we are unable to resist overeating when served large portions because much of our genetic code evolved over hundreds of thousands of years when our ancestors had good reason indeed to believe that tomorrow might never come, literally but also in terms of the next meal. "Eat while you can; you never know when you'll get another chance," our genes seem to be telling us. Your parents trained you to clean up your plate, and your genes demand it as well.

## That Thing Is Gonna Ruin Your Eyes

This was my parents' warning to their son who watched too much TV. I watched it plenty, and the TV didn't make me or my friends go blind, just as I doubt that video screens will blind any of today's kids. But my parents were onto something, perhaps without knowing exactly what. My generation, the baby boomers, was the first to grow up with television. Previous generations, up to and including that of my parents, didn't have a choice; they had to do something else with their eyes. My generation had a choice. Video screens were certainly abundant, but they had not yet taken over our visual world. Today, their domination is nearly complete.

I once heard Garrison Keillor compare surfing the Internet to going to the state fair. His point was that there really is no comparison. The sensory pleasures, and displeasures, of the fair, its sights, sounds, smells, tastes, and textures, are in a different league altogether from the sterile, two-dimensional world of the Internet, and Keillor lamented our apparent preference for the latter.

Consider all the video screens in your life. As the coffee percolates, the TV in the kitchen provides the morning news and weather. At the breakfast table, a teenager watches a movie on her iPhone, all the better to avoid making eye contact with her parents, and especially her siblings. Seated in the backseat of an SUV, two toddlers are anaesthetized by video cartoons emanating from the backs of the headrests in front of them. At work, we are so thoroughly yoked to our desktop computer monitors that we type electronic messages to colleagues seated at their desks in adjacent offices rather than walking a few steps to see them face-to-face. At the health club, we line up in rows on treadmills or stationary bikes, each fronted by its own video monitor. The advent of relatively low-cost plasma and LCD flat-screen monitors has made it much easier, both financially and in terms of interior design, for restaurants and bars to install TVs in so many different locations that in some of the trendier new establishments, it is nearly impossible to find a seat that is not directly in the line of sight of one. In late 2006, Walmart announced a pilot program wherein parents could rent, for a small fee, grocery carts with built-in video screens for their young children to watch while Mom or Dad shopped. Pedestrians, passengers, drivers, and students stare endlessly at the tiny touch screens of their wireless communication devices.

One of my more perceptive students recently explained to me that when he and his peers are around folks from my generation, they have learned not to constantly stare at their touch-screen wireless devices. "We know you guys don't like that," he said. When they're with each other, however, the hand-held gadgets play at least as important a role in their discourse as the flesh-and-blood members of the conversation. In social situations like this, there appears to be scarcely any distinction between persons present in the flesh and the digital participants. It has been said that the recent revolution in wireless communication represents the greatest generational upheaval since the 1960s. You won't get any arguments to the contrary from me.

## Noise Pollution

As I type these words, sitting at home in front of my computer, I pause to listen to my surroundings. The cooling fan inside the computer hums along, a constant whir of white noise. Is it loud enough to be considered a distraction? From above, I can hear as well as feel the rush of air through

# Chapter 10 • Perception and Education

## Observation

Nobel Prize winner Pierre-Gilles de Genne (1932–2007) noted that the invention process begins with observation. The keenest observers among us tend to be the most prolific inventors. They observe a need, a phenomenon, an incongruity in modern life. From these, invention flows. To the great inventors, perhaps observation comes naturally, but that is certainly no reason not to teach it. Make no mistake: the ability to *observe* is no longer taught or emphasized the way it once was. And observation, by definition, involves perception.

My students seem much less interested in improving their perceptions, such as their ability to observe in the lab, than in learning to use the sophisticated instruments designed to observe for them. They are happy to devote considerable effort to mastering these new technologies, to the detriment of the development of their own instruments, their senses. While modern instrumentation is wonderful, I find it disturbing that we no longer train our senses as we once did. Just as we nourish our bodies with a variety of foods, so our brains need to be nourished with a rich variety of sensory inputs.

Aside from laser eye surgery and cochlear implants, there's not much we can change to improve our sensory hardware, but we can, through experience and education, get better at perceiving. Not long ago, training

one's eyes was a vital part of the educational process. Depending on the era, where you grew up, and your social status, such training may have involved serious instruction in painting, drawing, and sculpture and endless hours learning to become proficient with the bow and arrow or the rifle. Discerning the difference between edible and poisonous plants was a matter of life or death, as was, for sailors, reading the sky to know when storms were coming.

None of this was really thought of as "training one's sense of vision." It was simply "learning a skill," or perhaps just "learning how to survive." Craftspeople of every persuasion, from woodworkers to stone masons to seamstresses, succeeded or failed according to how well trained their eyes were.

Over the years, I have found that engineers who can sketch very well are almost always also very perceptive visually. They visualize complex mechanisms in three dimensions, they notice subtle details in things, and they have a great sense of proportion, so that they can tell, for example, whether a modified component will fit in the space available. Modern computer-aided engineering software tools help the rest of us with these perceptions, but engineers who possess this ability are generally superior performers.

A colleague comes into an engineer's office and says, "Listen, when we install that new motor out in Unit 12, we're going to have to modify the structure around it, because the new motor is bigger than the old one." "Well, you might be right, but how?" He says, "Let's try this," and he grabs a sheet of paper and starts sketching. After a while, you stop him and say, "That looks good, but how about this idea—it might be cheaper and faster." You add a few lines to his sketch, and maybe erase a few others, to which he responds with some modifications of his own. The sketch gets passed back and forth several times like this, and soon a working conceptual design has been achieved. This is only the beginning of the project. In the next phase, one of you will be seated in front of a computer. But I cannot count the number of times I have found myself in this scenario as an engineer. This sort of communication through collaborative sketching

isn't formally taught very much in engineering school, although it should be. The best collaborative sketchers are those with the best visual perception. Artistic talent certainly plays a role, but so does training. We can train ourselves to see better—to be better observers of our visual world. And the younger the student, the better. Just don't stop the training when she gets to the third grade.

## Perception and the Arts

Our senses and perceptions keep us alive, physically, intellectually, and emotionally. They educate us. We marvel at the eyes of infants—practically everything they see is new. Their eyes are efficient, prolific data acquisition devices, dispatching information by the truckload to the brain in ways we don't perfectly understand. Language pours in through the ears. The child's brain is optimized for language learning in ways that educators have only recently begun to comprehend and take advantage of.

Not so long ago, perception was mainly a matter of survival. As living conditions improved, the arts developed, and they found their way into formal education. Nowadays, however, training in the arts is sometimes viewed as frivolous, irrelevant, or elitist. In my community, the schools with the best instruction in music and the visual arts are generally the most expensive private schools. At many public schools, where budgets are tight, arts programs tend to struggle for funding and respect. "Why should they teach my kid to play the violin, or to paint with water colors? That's not going to get him a job," many parents say. But in the long run, this type of education really could help that young student get a job—as a doctor, a lawyer, a teacher, an engineer, a salesperson, a police officer, a plumber, or an entrepreneur.

Even at the expensive private high school my stepkids attended, it is quite easy to graduate without ever taking a course in any of the arts. Yet each of my kids received a semester's training in "computer keyboarding." A semester of computer keyboarding, for today's kids, is about as pointless and redundant as a semester's instruction on "breathing" would be. Arts education is crucial not because it creates artists, or even because it fosters appreciation of the arts. Rather, arts education is important because it hones a person's perceptions. And that's how it should be taught.

## Learning Styles and Perception

In the behavioral sciences, it is widely believed that individuals have different "learning styles," or ways in which they approach learning tasks. Formal inquiries into learning styles date back to at least the 1970s. One of the more prominent learning-style theories, called either VAK or VARK, has its roots in the senses. VAK stands for Visual, Auditory, and Kinesthetic (the R in VARK is for Reading). Visual learners, the theory proposes, learn best through images, and they tend to think in pictures.[1] Auditory learners learn best through lectures and discussions, while kinesthetic learners tend to be hands-on learners, preferring experiments and projects. Reading-style learners prefer to learn through reading and writing.

Just about anyone who has taught long enough can tell you that there is something to this theory. Students really do seem to have preferences for learning through one sense or another. One problem with applying this theory to your teaching, though, is that, even in a small class, you're likely to have students spanning the range of learning styles. If you favor one teaching style, you will be less effective at reaching students who prefer a different style. My approach to the problem is to try to include visual, auditory, and kinesthetic elements in all of my interactions with students. When I lecture, I talk to the auditory learners; I draw lots of pictures on the board and show images through a computer projector for the visual learners; and I do a demonstration or pass around a show-and-tell item for the kinesthetic folks. I hope this gives everyone a chance to exercise his or

---

[1]In *What Do You Care What Other People Think?* Richard Feynman recounts a discussion between two young boys about the nature of thought. One says "Thinking is nothing but talking to yourself inside." The other boy then asks his friend what words he imagines when he thinks about the form of some oddly shaped object, like the crankshaft of an engine.

her preferred learning style while at the same time practicing the others. This varied approach also breaks things up nicely and tends to be more lively than a straight-up talking-head lecture. One of my aims is to reduce the number of students who think they learn best while sleeping.

In *What the Best College Teachers Do,* Ken Bain draws a similar conclusion. Bain quotes a teacher who notes that one positive thing that all the research into learning styles has done is to "call attention to the need to diversify. I don't think there's much evidence that most people have exclusive learning styles and can't learn in any way but one, but I do think that we all benefit from variety." Bain also quotes Vanderbilt's Jeanette Norden, who puts it more succinctly. "The brain loves diversity," she says.

## A Good Ear

You might not think tennis has anything in common with music, but there is at least one connection. The tension in the strings of a tennis racket is an important variable. The lower the tension in the strings of my racket, the harder I can hit the ball. So when I go out to play, I'm usually checking to see which of my rackets has the loosest strings that day. The tension in the strings on a racket, like the strings on a guitar, changes over time and with the weather, so I can't always be sure which racket has the loosest strings on any given day. Tennis players have a quick test for this, and that is to "ping" the strings of a racket by hitting them with the edge of another racket. A musical tone is produced, and the lower the tone, the looser the strings. It's easy for most players to judge string tension this way, but I quit doing it years ago, because I can almost never tell which racket has the loosest strings. When I ping racket A against racket B, and then racket B against racket A, sometimes they sound different, but I generally can't judge which tone is the lowest. The ping test just doesn't work for me.

What kind of idiot can't tell which of two musical tones has the lowest pitch? My kind, I guess. I have the opposite of what you call in music a "good ear." Having a good musical ear is an enormously complex combination of nature and nurture; that is to say, it is an intricate blending of sensation and perception. I'm really not sure about the nature of my musical ear, because there has been such an utter lack of nurture.

As a kid, I stubbornly resisted all parental attempts at musical training. We had a piano in our house, and although my mother taught my younger brother to play respectably well, I refused to go near that piano. Both parents cajoled me to join the church choir (Dad said it was a good way to meet girls), but I was having none of that, either. And then there was the incident with old Mrs. Townsend, my fourth-grade teacher. Every year at school, she produced a musical play with a cast of students. Unfortunately, each and every student, willing or not, was required to audition for a singing role in the play. To say that my brief audition, singing a few scant lines accompanied by Mrs. Townsend at the piano and witnessed by an audience of my classmates, didn't go well would be an understatement. To no one's surprise, I didn't get a singing role in the play. I imagine no one else thought twice about my little musical debut, but I have never forgotten how intensely and awfully self-conscious it made me feel to have to sing in front of that class. I resolved never to feel that way again. And I haven't.

But music is, or should be, an important part of a child's education. By avoiding any and all musical training, I shortchanged myself; there is no question in my mind. Does this explain why I can't judge the relative tension in the strings in my tennis rackets when I ping them?

## Amusia

In spite of all this, music is an important part of my life. I love listening to lots of different kinds of music, I attend concerts regularly, and I own a relatively large collection of recorded music by my favorite artists. I may not have much musical talent, or training, or a good ear, but I don't believe that my musical difficulties rise to the level of amusia. *Amusia* is a term that describes a variety of conditions related to the inability to recognize various aspects of music, such as tones, melodies, or rhythms. Amusia can be congenital, or it can develop as a result of a brain injury or disease. Oliver Sacks describes the case of a woman he calls D.L. in *Musicophilia*. D.L.

described music, any music, as sounding to her like pots and pans being thrown onto the kitchen floor. At public events, she could not recognize "The Star-Spangled Banner" and only knew to stand when others around her did so. She couldn't identify "Happy Birthday," even though, as a teacher, she played a recording of it in her classroom every time a student had a birthday. Now, that's amusia. D.L. was an intelligent woman, and her hearing was normal—there was nothing wrong with her ears. She recognized people's voices normally and had no difficulty distinguishing everyday sounds such as car horns, barking dogs, and running water.

I don't think I've ever met anyone quite like D.L., although I could have; her condition is not the sort of thing people advertise about themselves. Some well-known individuals have suffered from amusia. Ulysses S. Grant reported that he knew only two songs. "One is Yankee Doodle," he said, "and the other isn't." And Vladimir Nabokov writes this in his autobiography *Speak, Memory:* "Music, I regret to say, affects me merely as an arbitrary succession of more or less irritating sounds."

Amusia was described in the medical literature as far back as 1878, but it received relatively little study until the past few decades. Someone like D.L. has extreme difficulty in discriminating music based on both pitch and timbre. Timbre relates to the quality or richness of any sound, not just music. Timbre is influenced by acoustic phenomena such as overtones and harmonics and is what makes the same note played on two different musical instruments sound different. Perceiving timbre is likewise complex. It may have something in common with the perception of color; coincidentally or not, timbre is sometimes described as "tone color" or in other color-related terms.

How could someone like D.L., with all her timbre-related difficulties, possibly be able to distinguish people based on the sounds of their voices? When someone says "hello" on the telephone, the characteristics we use to identify the person are mostly a matter of timbre. And yet D.L. *could* tell people apart based on their voices. It turns out that voice-selective areas in the brain are probably separate from the areas that perceive musical timbre. In light of our earlier discussions of face blindness and other aspects

Perception

of area-specific specialization within the brain, that's not particularly surprising.

## Taste Testing

As an engineering professor, I often get to tour factories and plants. I've visited all kinds, from huge, complex places like nuclear power plants, airplane manufacturing facilities, and automobile assembly lines all the way down to a small factory in France where they make the little wire gizmos (called *muselets*) that hold the cork inside a champagne bottle. Food manufacturing plants are some of the most interesting places to visit. The modern food plant is a fascinating blend of high-tech engineering prowess and old-fashioned, even prehistoric human perception. Precise instrumentation, computer controls, modern materials, and antiseptic conditions are the norm in the modern food factory. But the taste tester, whose job has scarcely changed over the centuries, remains.

I've met taste testers in pastry factories, ice cream plants, and whiskey distilleries, among other places. Whiskey taste testers do their job (sober, thank you) by applying the product to their tongues a drop at a time. Like taste testers in food factories, they spend months in training and, once certified, often have the power to reject an entire batch of product. I've even heard that dog-food plants employ human taste testers, since dogs' evaluations of the tastes and smells of their meals are hard for humans to understand with precision. Food-plant managers have told me that finding good candidate taste testers and then developing them is often quite difficult, and not just in dog-food plants. Most of our palates simply aren't discriminating enough, even after the rigorous apprenticeships the trainees undergo.

## In the Mood

It's easy to bemoan the rough treatment our senses receive in the world we have created, one crammed with sterile video screens, noise pollution, noxious odors, and too-large portions of overly processed, flavorless foods. All the same, we have a lot more control over our senses than we might realize.

Jill Taylor, the brain scientist we met earlier, suffered a massive stroke from which it took eight years to fully recover. In *My Stroke of Insight,* she gives a lot of advice that relates to our senses. Taylor knows what it is like to have great chunks of her perceptive software rendered inoperative. Those perceptive abilities, even when wrenched away by something as violent as a stroke, can sometimes be recovered. Much of her book discusses how her experiences recovering from her stroke have put her more in touch with her moods and emotions, and thus with how to control her feelings.

Taylor offers suggestions that can help us all get better at using our senses, regardless of our age or condition. And the senses are wonderful mood modifiers. Music? Of course. And color. But what about smell? Candles, incense, sautéed onions. If I want my wife or one of the kids to walk into the kitchen and say, "Mmm, it smells good in here." all I have to do is sauté some onions. It works every time.

Taylor is convinced that almost everyone can improve her sense of smell, and my limited experience tends to agree. Every year I include an experiment in one of my courses wherein the students perform tests on different kinds of plastic. In one test, they burn a small piece of plastic, extinguish the fire, and then carefully sniff the piece. Different plastics have different characteristic smells that depend on their chemical makeup. Polyethylene is one of the easiest to identify. It smells just like candle wax. I've watched students take a little sniff of burned polyethylene, think for a second, then spontaneously burst into "Happy Birthday." Smell is like that. But some of the plastics have smells that are much more difficult to identify. And many of my students can't identify any of the burned plastics, even polyethylene, beyond "It smells like burned plastic." This would be a little like walking into the kitchen and saying, "It smells like food is cooking in here."

I used to think there was nothing to be done about it. Some of my students can identify the smells, while others can't. But Jill Taylor is right;

253

with a little practice, we can get better at smelling. Nowadays, so that students can train their noses, I expose them to labeled containers of various scents, such as cinnamon, vinegar, and apple juice, which correspond to certain types of plastics. Although we don't spend much time on this, it seems to work. Many of my students go from "it smells like burned plastic" to a much more useful conclusion after just a little training.

So there's a good chance you can get better at smelling things, a skill that is useful for other applications besides identifying plastics. You can likewise refine your senses of taste, hearing, and sight. Maybe it will make you a better cook, musician, hunter, or painter. And maybe not. There are plenty of other reasons to make the effort. A few years back, the Mozart Effect made a big splash in the popular press, and playing Mozart to babies in the womb, or to students getting ready to take an important test, was all the rage. The idea was that listening to Mozart's music was supposed to make you smarter. Heaven forbid we should listen to Mozart because the music is beautiful. Listening to Mozart because it might make you smarter is letting your left brain rule your life. Give your right brain a chance. Relax, close your eyes, and hear the music.

# Bibliography

## Introduction

Aristotle. *De anima.* New Haven, CT: Yale University Press, 1959.

Bruemmer, Fred. *The Narwhal.* Shrewsbury, UK: Swan Hill Press, 1993.

Clarke, Arthur C. *Profiles of the Future.* New York: Henry Holt, 1984.

Edwards, Betty. *Drawing on the Right Side of the Brain.* Los Angeles: J. P. Tarcher, 1979.

Ferdinand, Pamela. "A Flexible, 9 Ft. Whale Tooth with Super Sensing Power?" http://news.nationalgeographic.com/news/2005/12/1213_051213_narwhal_tooth.html, December 13, 2005, accessed February 22, 2011.

Frisch, Karl von. *Bees: Their Vision, Chemical Senses, and Language.* London: Jonathan Cape, 1968.

Hunt, Frederick Vinton. *Origins in Acoustics: The Science of Sound from Antiquity to the Age of Newton.* New Haven, CT: Yale University Press, 1978. Quote at p. 11.

Kreiser, John. "A Teen Who Sees with Sound." www.cbsnews.com/stories/2006/09/06/eveningnews/main1977730, accessed February 22, 2011.

Yantis, Steven, ed. *Steven's Handbook of Experimental Psychology.* Vol. 1, *Sense and Perception.* 3rd ed. New York: John Wiley, 2002.

## Part 1. Stimulus

### Chapter 1. Electromagnetic Stimuli

Al-Haytham, Ibn. *Optics of Ibn Al-Haytham: Books 1–3.* London: Warburg Institute, 1989.

ASTM G173-03. Standard Tables for Reference Solar Spectral Irradiances, 2008. American Society for Testing and Materials, West Conshohocken, PA.

Fisher, David E., and Marshall J. Fisher. "The Color War." *American Heritage of Invention and Technology* 12, no. 3 (1997): 8–18.

Hecht, Jeff. "How We Became Wired with Glass." *American Heritage of Invention and Technology* 15, no. 3 (2000): 44–53.

Herbert, Robert L., and Neil Harris. *Seurat and the Making of "La Grande Jatte."* Berkeley: University of California Press, 2004.

Layton, Julia. "How Remote Control Works." http://electronics.howstuffworks.com/remote-control2.htm, accessed March 15, 2011.

"Midnight Zone." www.extremescience.com/zoom/index.php/ocean-zones/94-midnight-zone, accessed February 22, 2011.

Nassau, Kurt. *The Physics and Chemistry of Color.* 2nd ed. New York: John Wiley, 2001. Quote at p. 8.

Perkowitz, Sidney. *Empire of Light: A History of Discovery in Science and Art.* Washington, DC: Joseph Henry Press, 1996.

Sennebogen, Emilie. "How Can a Machine Match a Paint Color Perfectly?" http://tlc.howstuffworks.com/home/machine-match-paint.htm, accessed March 15, 2011.

Underhill, Paco. *Call of the Mall.* New York: Simon and Schuster, 2004.

———. *Why We Buy: The Science of Shopping.* New York: Simon and Schuster, 1999.

Warren, Richard, and Roslyn Warren. *Helmholtz on Perception: Its Physiology and Development.* New York: John Wiley, 1968.

## *Chapter 2.* Chemical Stimuli

Burr, Chandler. *The Emperor of Scent: A True Story of Perfume and Obsession.* New York: Random House, 2002.

Holmes, Oliver Wendell. *The Autocrat of the Breakfast-Table.* Pleasantville, NY: Akadine Press, 2001. Quote at p. 266.

"How Is Pepper Heat Measured?" www.eatmorechiles.com/Scoville_Heat.html, accessed March 14, 2011.

Lanchester, John. "Scents and Sensibility." *New Yorker,* March 10, 2008, 120–22.

Russell, Bertrand. *A History of Western Philosophy.* New York: Simon and Schuster, 1972.

Scoville, Wilbur L. "Notes on Capsicums." *Journal of the American Pharmacists Association* 1 (1912): 453–54.

Stoneham, Marshall. "Making Sense of Scent." *Materials Today* 10, no. 5 (May 2007): 64.

Turin, Luca. "A Spectroscopic Mechanism for Primary Olfactory Reception." *Chemical Senses* 21 (1996): 773–91.

Turin, Luca, and Tania Sanchez. *Perfumes: The Guide.* New York: Viking, 2008. Quote at p. 197.

## Chapter 3. Mechanical Stimuli

Bibel, George. *Beyond the Black Box: The Forensics of Airplane Crashes.* Baltimore: Johns Hopkins University Press, 2007.

Katz, David. *The World of Touch.* London: Psychology Press, 1989.

Needham, Joseph, and Wang Ling. *Science and Civilisation in China.* Vol. 4, Part 2. Cambridge: Cambridge University Press, 1965.

Rossing, Thomas D. *The Science of Sound.* 2nd ed. Reading, MA: Addison-Wesley, 1990.

Sacks, Oliver. *The Man Who Mistook His Wife for a Hat.* New York: Summit Books, 1970.

Stevens, Stanley S. *Psychophysics: Introduction to Its Perceptual, Neural, and Social Prospects.* New Brunswick, NJ: Transaction Books, 1986.

Stevens, Stanley S., and F. Warshofsky. *Sound and Hearing.* New York: Time, 1965.

Yantis, Steven, ed. *Steven's Handbook of Experimental Psychology.* Vol. 1, *Sensation and Perception.* 3rd ed. New York: John Wiley, 2002.

## Chapter 4. The Science of Sensation

Fechner, Gustav. *Elements of Psychophysics.* New York: Holt, Rinehart, and Winston, 1966.

Stevens, Stanley S. *Psychophysics: Introduction to Its Perceptual, Neural, and Social Prospects.* New Brunswick, NJ: Transaction Books, 1986. Quotes at p. 12.

# Part 2. Sensation

Chorost, Michael. *Rebuilt: My Journey Back to the Hearing World.* Boston: Houghton-Mifflin, 2005.

## Chapter 5. Vision

Atchison, David A., and George Smith. *Optics of the Human Eye.* Oxford: Butterworth-Heinemann, 2000.

Baylor, D. A., T. D. Lamb, and K. W. Yau. "Responses of Retinal Rods to Single Photons." *Journal of Physiology* 288 (1979): 613–34.

Darwin, Charles. "Difficulties of the Theory." In *The Origin of Species.* Madison, WI: Cricket House Books, 2010.

*The Eye Digest.* www.agingeye.net/visionbasics/theagingeye.php, accessed March 15, 2011.

Fernald, R. D. "Evolution of Eyes." *Current Opinion in Neurobiology* 10 (2000): 444–50.

Figueiro, Mariana G. "Lighting the Way." www.lrc.rpi.edu/programs/lightHealth /AARP/pdf/AARPbook1.pdf, accessed March 15, 2011.

Hofer, H., B. Singer, and D. R. Williams. "Different Sensations from Cones with the Same Photopigment." *Journal of Vision* 5 (2005): 444.

Jackson, G. R., and C. Owsley. "Scotopic Sensitivity during Adulthood." *Vision Research* 40, no. 18 (2000): 2467–73.

Javal, Emile. *Physiologie de la lecture et de l'écriture.* Paris: Félix Alcan, 1905.

Koch, Kristin, Judith McLean, Ronen Segev, Michael Freed, Michael Berry, Vijay Balasubramanian, and Peter Stirling. "How Much the Eye Tells the Brain." *Current Biology* 16, no. 14 (July 25, 2006): 1428–34.

Kolb, Helga, Eduardo Fernandez, and Ralph Nelson. "The Organization of the Retina and Visual System." http://webvision.med.utah.edu/, accessed May 23, 2011.

Krader, Cheryl G. "Artificial Corneas." *Eurotimes* 16, no. 2 (2011): 12.

Land, M. F., and R. D. Fernald. "The Evolution of Eyes." *Annual Review of Neuroscience* 15 (1992): 1–29.

Nassau, Kurt. *The Physics and Chemistry of Color.* 2nd ed. New York: John Wiley, 2001.

Nathans, Jeremy. "The Evolution and Physiology of Human Color Vision: Insights from Molecular Genetic Studies of Visual Pigments." *Neuron* 24 (October 1999): 299–312.

National Eye Institute. www.nei.nih.gov/health/maculardegen/armd_facts.asp#1a, accessed March 14, 2011.

Newport, John Paul. "Golf Journal: The Eyes Have It." *Wall Street Journal,* October 27, 2007.

Nolte, John. *The Human Brain: An Introduction to Its Functional Anatomy.* 6th ed. Philadelphia: Mosby Elsevier, 2009. Quote at p. 432.

Perkowitz, Sidney. *Empire of Light: A History of Discovery in Science and Art.* Washington, DC: Joseph Henry Press, 1996.

Rosetti, Hazel. *Colour.* Princeton, NJ: Princeton University Press, 1983.

Schneeweis, D. M., and J. L. Schnapf. "Photovoltage of Rods and Cones in the Macaque Retina." *Science* 268 (1995): 1053–56.

Sichert, A. B., P. Friedel, and J. L. van Hemmen. "Snake's Perspective on Heat: Reconstruction of Input Using an Imperfect Detection System." *Physical Review Letters* 97 (2006): 068105-1 to 068105-4.

Stevens, Stanley S. *Psychophysics: Introduction to Its Perceptual, Neural, and Social Prospects*. New Brunswick, NJ: Transaction Books, 1986.

Underhill, Paco. *Call of the Mall*. New York: Simon and Schuster, 2004.

## Chapter 6. The Chemical Senses

Ackerman, Diane. *A Natural History of the Senses*. New York: Vintage Books, 1990.

Avicenna. *The Canon of Medicine*. Chicago: Kazi, 1999.

Bartoshuk, Linda M. "Sweetness: History, Preference, and Genetic Variability." *Food Technology* 45, no. 11 (1991): 112–13.

Bilger, Burkhard. "The Search for Sweet." *New Yorker,* May 22, 2006, 40–46.

Buck, Linda, and Richard Axel. "A Novel Multigene Family May Encode Odorant Receptors: A Molecular Basis for Odor Recognition." *Cell* 65, no. 1 (April 5, 1991): 175–87.

Burr, Chandler. *The Emperor of Scent: A True Story of Perfume and Obsession*. New York: Random House, 2002.

Corey, David, and Charles Zuker. "Sensory Systems." *Current Opinion in Neurobiology* 6 (1996): 437–39.

Green, B. G. "Referred Thermal Sensations: Warmth versus Cold." *Perception and Psychophysics* 22 (1977): 331.

McClintock, Martha K. "Menstrual Synchrony and Suppression." *Nature* 229, no. 5282 (1971): 244–45.

Nolte, John. *The Human Brain: An Introduction to Its Functional Anatomy*. 6th ed. Philadelphia: Mosby Elsevier, 2009.

Sacks, Oliver. *The Man Who Mistook His Wife for a Hat*. New York: Summit Books, 1970. Quotes at pp. 149, 150–51.

Schwartz, John. "Picked from a Lineup, on a Whiff of Evidence." *New York Times,* November 4, 2009.

Senomyx. www.senomyx.com/flavor_programs/receptorTech.htm, accessed August 3, 2010.

Settles, Gary S. "Sniffers: Fluid-Dynamic Sampling for Olfactory Trace Detection in Nature and Homeland Security: The 2004 Freeman Scholar Lecture." *Journal of Fluids Engineering* 127, no. 2 (2005): 189–218.

Sulloway, Frank J. *Freud, Biologist of the Mind*. New York: Basic Books, 1979.

Thomas, Lewis. "On Smell." *New England Journal of Medicine,* March 27, 1980, 731–33.

Vroon, Piet. *Smell: The Secret Seducer*. New York: Farrar, Straus and Giroux, 1997.

Woolf, Harry, ed. *Quantification: A History of the Meaning of Measurement in the Natural and Social Sciences.* Indianapolis: Bobbs-Merrill, 1961. Hippocrates quote at p. 89.

Zhao, Grace Q., Yifeng Zhang, Mark Hoon, Jayaram Chandrashekar, Isolde Erlenbach, Nicholas Ryba, and Charles Zuker. "The Receptors for Mammalian Sweet and Umami Taste." *Cell* 115 (2003): 255–66.

Zucchino, David. "Bomb-Sniffing Dogs Are Soldiers' Best Friends." *Los Angeles Times,* July 24, 2010, http://articles.latimes.com/2010/jul/24/world/la-fg-afghani stan-dogs-20100725, accessed July 28, 2010.

## *Chapter 7.* The Mechanical Senses

Allin, E. F. "Evolution of the Mammalian Middle Ear." *Journal of Morphology* 147, no. 4 (1975): 403–37.

Asimov, Isaac. *The Human Body: Its Structure and Operation.* Boston: Houghton Mifflin, 1963. Quote at p. 226.

Békésy, Georg von. *Experiments in Hearing.* New York: McGraw-Hill, 1960.

Bronzaft, Arline L., and Dennis P. McCarthy. "The Effect of Elevated Train Noise on Reading Ability." *Environment and Behavior* 7 (December 1975): 517–28.

Chorost, Michael. *Rebuilt: My Journey Back to the Hearing World.* Boston: Houghton-Mifflin, 2005. Quote at p. 203

Cole, Jonathon D. *Pride and a Daily Marathon.* Cambridge, MA: MIT Press, 1995.

Craig, James C., and Gary B. Rollman. "Somesthesis." *Annual Review of Psychology* 50 (1999): 305–31.

Eliot, Lise. *What's Going On in There? How the Brain and Mind Develop in the First Five Years of Life.* New York: Bantam Books, 1999.

Fay, R. R. *Hearing in Vertebrates: A Psychophysics Databook.* Winnetka, IL: Hill-Fay Associates, 1988.

Fletcher, Dan. "How High Can I Crank My iPod's Volume?" www.time.com/time /business/article/0,8599,1926796,00.html, September 30, 2009, accessed March 15, 2011.

Hain, Timothy C. "Head Impulse Test and Head Heave Test." www.dizziness-and -balance.com/practice/head-impulse.html, accessed March 15, 2011.

Heath, Thomas L., ed. *The Works of Archimedes with the Method of Archimedes.* New York: Dover, 1953. Quote at p. xix.

Heathcote, J. A. "Why Do Old Men Have Big Ears?" *British Medical Journal* 311 (1995): 1668.

"High Speed Robot Hand." www.ebaumsworld.com/video/watch/80731612/, accessed March 12, 2011.

"History of Gyroscopes." www.gyroscopes.org/history.asp, accessed March 14, 2011.

Hunt, Frederick Vinton. *Origins in Acoustics: The Science of Sound from Antiquity to the Age of Newton.* New Haven, CT: Yale University Press, 1978.

Keller, Helen. *Helen Keller in Scotland: A Personal Record Written by Herself.* London: Methuen, 1933. Quote at p. 68.

Nolte, John. *The Human Brain: An Introduction to Its Functional Anatomy.* 6th ed. Philadelphia: Mosby Elsevier, 2009.

Pelz, L., and B. Stein. "Zur klinischen Beurteilung der Ohrgröße bei Kindern und Jugendlichen." *Pädiatrie und Grenzgebiete* 29 (1990): 229–35.

Robles-De-La-Torre, Gabriel. "The Importance of the Sense of Touch in Virtual and Real Environments." *IEEE Multimedia,* July–September 2006, 24–30.

Schopenhauer, Arthur. "On Noise." In *Complete Essays of Schopenhauer: Seven Books in One Volume,* book 5. Translated by T. Bailey Saunders. New York: Wiley, 1942.

Stevens, Stanley S. *Hearing: Its Psychology and Physiology.* New York: John Wiley, 1938.

Stevens, Stanley S., and Fred Warshofsky. *Sound and Hearing.* New York: Time, 1965.

Vallbo, A. B., and R. S. Johansson. "Properties of Cutaneous Mechanoreceptors in the Human Hand Related to Touch Sensation." *Human Neurobiology* 3 (1984): 3–14.

Wang, Shirley S. "Can a Tiny Fish Save Your Ears?" *Wall Street Journal,* August 4, 2009.

## Part 3: Perception

### Chapter 8. Remembering the Present

Ackerman, Diane. *A Natural History of the Senses.* New York: Vintage Books, 1995.

Allen, Grant. "Note-Deafness." *Mind* 10 (1878): 157–67.

Blount, Roy. *Alphabet Juice.* New York: Sarah Crichton Books, 2008.

Brodmann, K. *Vergleichende Lokalisation hehre der Grosshirnrinde in ihren Prinzipien dargestellt auf Grund des Zellenbaues.* Leipzig: J. A. Barth, 1909.

Bronzino, Joseph D., ed. *Biomedical Engineering Fundamentals.* Boca Raton, FL: Taylor and Francis, 2006.

Chorost, Michael. *Rebuilt: My Journey Back to the Hearing World.* Boston: Houghton-Mifflin, 2005. Quote at p. 87.

Cytowic, Richard E. *The Man Who Tasted Shapes.* Cambridge, MA: MIT Press, 2003.

————. *Synesthesia: A Union of the Senses.* 2nd ed. Cambridge, MA: MIT Press, 2002.

Degenaar, Marjolein. *Molyneux's Problem: Three Centuries of Discussion on the Perception of Forms.* Boston: Kluwer, 1996.

Djourno, André, and Charles H. Eyriès. "Prothèse auditive par excitation électrique à distance du nerf sensoriel à l'aide d'un bobinage inclus à demeure." *La Presse Medicale* 65 (1957): 1417.

Duffy, Patricia L. *Blue Cats and Chartreuse Kittens: How Synesthetes Color Their Worlds.* New York: Henry Holt, 2002. Quote at p. 1.

Edelman, Gerald. *The Remembered Present: A Biological Theory of Consciousness.* New York: Basic Books, 1989.

Engel, Howard. *The Man Who Forgot How to Read: A Memoir.* New York: Thomas Dunne Books, 2007.

Feynman, Richard. *What Do You Care What Other People Think?* London: W. W. Norton, 2001.

Gallagher, Shaun. *How the Body Shapes the Mind.* Oxford: Clarendon Press, 2005.

Galton, Francis *Inquiries into Human Faculty and Its Development.* London: J. M. Dent, 1883.

Gladstones, W. H., M. A. Regan, and R. B. Lee. "Division of Attention: The Single-Channel Hypothesis Revisited." *Quarterly Journal of Experimental Psychology: Human Experimental Psychology* 41 (A) (1989): 1–17.

Gregory, Richard L. *Eye and Brain: The Psychology of Seeing.* New York: McGraw-Hill, 1966. Quote at p. 198.

Gross, Charles G. "Genealogy of the 'Grandmother Cell.'" *Neuroscientist* 8, no. 5 (2002): 512–18.

"L-1 Identity Solutions." www.l1id.com, accessed March 15, 2011.

Levitin, Daniel J. *This Is Your Brain on Music.* New York: Dutton, 2006.

Mithen, Steven. "The Diva Within." *New Scientist,* February 23, 2008, 38–39.

Nolte, John. *The Human Brain: An Introduction to Its Functional Anatomy.* 6th ed. Philadelphia: Mosby Elsevier, 2009.

Ridley, Matt. *Genome: The Autobiography of a Species in 23 Chapters.* New York: HarperCollins, 2000.

Sacks, Oliver. "Face-Blind." *New Yorker,* August 30, 2010, 36–43.

————. "A Man of Letters." *New Yorker,* June 28, 2010, 22–28.

————. *The Man Who Mistook His Wife for a Hat.* New York: Summit Books, 1985. Quotes at pp. 12–13.

————. *Musicophilia.* New York: Vintage, 2007.

Syrus, Publilius. *The Moral Sayings of Publius Syrus, a Roman Slave.* Translated by D. Lyman. Cleveland: L. E. Barnard, 1856. Quote at p. 13.

Taylor, Jill Bolte. *My Stroke of Insight: A Brain Scientist's Personal Journey.* New York: Plume Books, 2006. Quotes at pp. 102–3.

Thompson, Peter. "Margaret Thatcher—A New Illusion." *Perception* 9 (1980): 483–84.

Warren, Richard M. "Perceptual Restoration of Missing Speech Sounds." *Science,* January 23, 1970, 392–93.

Wiesel, T. N., and D. H. Hubel. "Single-Cell Responses in Striate Cortex of Kittens Deprived of Vision in One Eye." *Journal of Neurophysiology* 26 (1963): 1003–17.

Zhao, W., R. Chellappa, A. Rosenfeld, and P. J. Phillips. "Face Recognition: A Literature Survey." *ACM Computing Surveys* (2003): 399–458.

## Chapter 9. Perception and Culture

Pollan, Michael. *In Defense of Food: An Eater's Manifesto.* London: Penguin Books, 2008. Quotes at pp. 54–55.

Schopenhauer, Arthur. "On Noise." In *Complete Essays of Schopenhauer: Seven Books in One Volume,* book 5. Translated by T. Bailey Saunders. New York: Wiley, 1942. Quote at p. 95.

## Chapter 10. Perception and Education

Bain, Ken. *What the Best College Teachers Do.* Cambridge, MA: Harvard University Press, 2004. Quotes at pp. 116–17.

Campbell, Don. *The Mozart Effect.* New York: Harper Paperbacks, 2001.

Feynman, Richard. *What Do You Care What Other People Think?* London: W. W. Norton, 2001. Quote at p. 54.

Fleming, Neil, and Colleen Mills. "Not Another Inventory, Rather a Catalyst for Reflection." *National Teaching and Learning Forum* 4 (1998): 137–49.

Sacks, Oliver. *Musicophilia.* New York: Vintage, 2007. Quotes at pp. 109 (Grant), 110 (Nabokov).

Taylor, Jill Bolte. *My Stroke of Insight: A Brain Scientist's Personal Journey.* New York: Plume Books, 2006.

# Index